Wastewater Engineering

and Treatment 2

Wastewater Engineering and Treatment 2

Membranes, Hybrid Systems, Bacteria Beds and Biofilters, Biological Disks

Kader Gaid

ISTE

WILEY

First published 2026 in Great Britain and the United States by ISTE Ltd and John Wiley & Sons, Inc.

ISTE Ltd
27-37 St George's Road
London SW19 4EU
UK

www.iste.co.uk

John Wiley & Sons, Inc.
111 River Street
Hoboken, NJ 07030
USA

www.wiley.com

Library of Congress Control Number: 2026934969

British Library Cataloguing-in-Publication Data
A CIP record for this book is available from the British Library
ISBN 978-1-83669-056-6

The manufacturer's authorized representative according to the EU General Product Safety Regulation is Wiley-VCH GmbH, Boschstr. 12, 69469 Weinheim, Germany, e-mail: Product_Safety@wiley.com.

Contents

1

Membrane Bioreactors

1.1. Introduction

Membrane bioreactors (MBRs) provide biological treatment using activated sludge and membrane filtration. Due to membrane integration, treatment can be carried out without a clarifier. The presence of membranes in the process helps to avoid all the problems related to final clarification and increases the compactness of the treatment system.

The joint use of membranes and activated sludge emerged in the late 1960s. The technique has significantly evolved with the nature of the membranes used, and its application has considerably increased in the past 10 years. In terms of design, the MBR process is equivalent to the standard activated sludge process, but the secondary settling tank (clarification tank) is replaced by membrane filtration. Indeed, in the MBR process, membranes are barriers that retain all micro-particles, large molecules, viruses and bacteria but, at the same time, organic matter is biodegraded in a bioreactor. Compared to other standard biological systems, MBRs provide higher quality effluent thanks to higher particle retention (total liquid/solid separation) and a high biomass content (controlled by the sludge retention time independently of the hydraulic retention time) adapted to the pollutant load. However, the factors that influence MBR purification performances remain the properties of membranes, the membrane module structure, the operating conditions and the bioreactor parameters.

Therefore, biological treatment is possible at high sludge concentrations (MES concentration between 8 and 15 $g.L^{-1}$), which enables work with high volume loading with low mass loading.

Membrane treatment provides high reliability due to the physical barrier represented by the membrane: no sludge loss and very high quality treated water. This is why MBRs have many advantages compared to the conventional wastewater treatment processes. The result is small dimensions, high quality filtered water, good disinfection capacity, higher volume loading and less sludge production. Consequently, the MBR process became an interesting option for the treatment and reuse of industrial and municipal wastewaters.

The range of membrane filtration adapted to membrane bioreactors extends from microfiltration to ultrafiltration (between 0.1 and 0.02 μm).

The MBR process is more resilient to loading variations and ensures constant quality for the effluent exiting the treatment line. MBRs provide a purification installation performing two functions on an ongoing basis: a biological purification function and a clarification function. Hence, the removal of dissolved and particulate pollutants leads to obtaining treated water of excellent quality that can be used for a wide range of applications. MBRs are therefore a hybrid technology that combine a biological system and a stage of separation on a porous membrane. This technology is an improvement of the biological system with activated sludge dating back to 1905, characterized by the replacement of the standard secondary settling tank by a membrane filtration unit, whose remarkable selectivity represents a barrier that cannot be crossed by purifying species, irrespective of their flocculation state. The metabolization process with its two aspects (anabolism and catabolism) observed in a system of activated sludge is similar to the one involved in MBR.

The modules containing membranes can be placed either on the external loop with activated sludge or directly in the reaction tank.

An MBR operation involves several automatically unfolding stages, mainly linked to membrane filtration, whose result is treated water and membrane cleaning.

MBR performances are the result of a combination of three factors:

– characteristics of activated sludge;

– membrane properties;

– membrane operating conditions.

1.2. MBR configurations

The use of membranes instead of a clarifier changes the definition of the "soluble" part of pollution as, depending on the membrane retention threshold, membrane walls remain impenetrable for a fraction of – or even all – colloids, which leads to a significant increase in the COD reduction in the case of high COD in raw water compared to the conventional activated sludge process.

Furthermore, the surface of the membrane is calculated based on the hydraulic loading of the plant. The higher the latter, the wider the membrane surface should be.

Due to these two aspects, the MBR process is a highly competitive system for the treatment of highly concentrated wastewater.

However, the MBR filtration performances inevitably decrease with filtration time. This is due to the deposit of soluble and particulate materials on and in the membrane, resulting from the interactions between the activated sludge components and the membrane.

MBR implementation includes pretreatments and activated sludge-based biological treatment, with sludge extraction towards a treatment workshop in a membrane filtration facility with an associated reactive facility for membrane cleaning.

Main treatment domains	**MBR application domains**
Treatment of carbon pollution	Concentrated effluents
Nitrification	Industrial effluents: particularly competitive for effluents with a very high COD concentration ($\geq$ 2,000 mg.L^{-1})
Nitrification/denitrification	Rehabilitation of existing plants
Treatment of carbon, nitrogen and phosphorus	Severe implementation constraints
Disinfection, mainly related to bacteria, protozoa	Very high quality water is needed: in situ recycling of treated water, reuse of water for irrigation or industrial needs, infiltration of treated water, discharge in sensitive area
Removal of ions and micropollutants	Pretreatment for secondary filtration of the reverse osmosis type

Table 1.1. *Main fields of treatment and application of the MBR process*

Membranes are often arranged into “rows” or “series”, sets using shared equipment (permeate piping, cleaning, concentrate piping). The module configuration strongly influences the global performances of the MBR processes. This configuration must facilitate a homogeneous distribution of the flow rate, high turbulences at the surface of the membrane, low energy cost, a strategy for easy cleaning and easy maintenance.

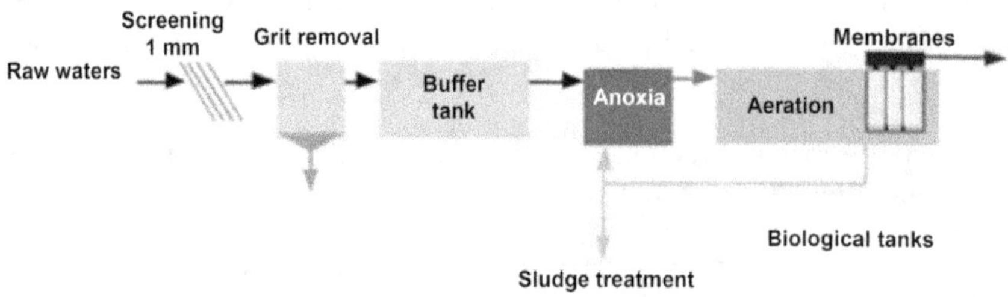

Figure 1.1. *Standard MBR wastewater treatment chain*

Considering these aspects, two general trends are manifest in practice.

Immersed membrane systems initially operated under cross-flow conditions. Depending on the concept, activated sludge could be pumped in the membrane modules placed next to the biological tank, leading to high performances and fluxes, but with high energy consumption.

Consequently, this was the preferred solution for the difficult cases of wastewater and for low flow rates. But due to airlift development and energy saving processes, it was extended to large volumes to be treated.

Two solutions are therefore implemented depending on membrane arrangement and insertion:

– a configuration known as immersed in a single tank (membranes immersed in the aerobic tank) (see Figure 1.2);

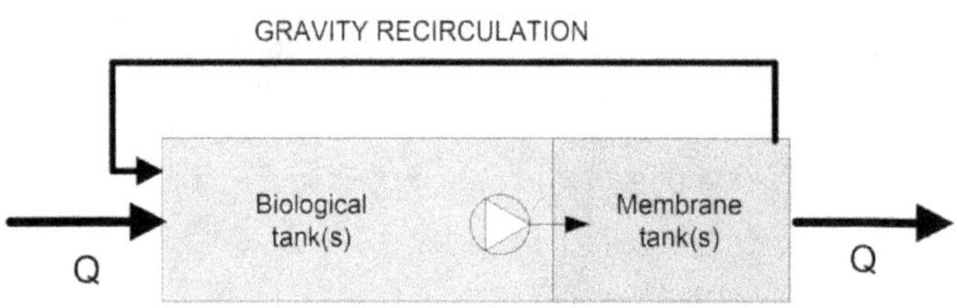

Figure 1.2. *Configuration known as immersed in a single tank*

– a configuration known as immersed in a separate tank (membranes immersed in tanks designed for filtration) (see Figure 1.3).

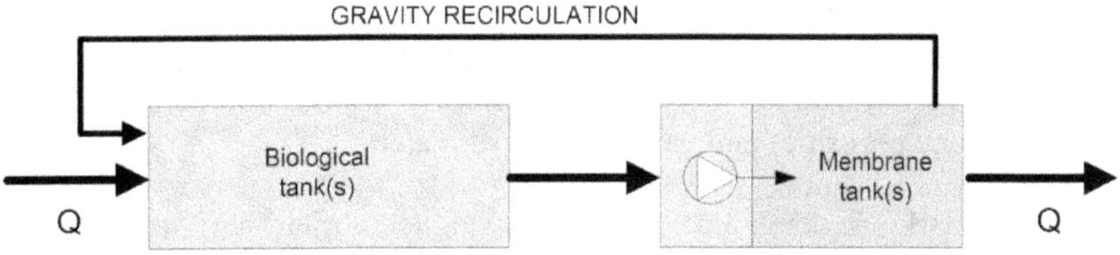

Figure 1.3. *Configuration known as immersed in a separate tank*

The membranes immersed in a single tank with the biological sludge are generally designed with hollow fiber or flat sheet membranes. A part of the air used to maintain aerobic conditions in the tank is then also used to support turbulences and proper circulation around membrane modules.

Due to the low turbulence facilitated around immersed membranes, they are preferred for the treatment of municipal wastewater.

Given their lower energy consumption, they are preferred by manufacturers. This system requires, however, intensive pretreatment and regular maintenance cleaning to maintain stable performances and avoid operational difficulties such as fiber clogging caused by hair or a poor distribution of turbulences.

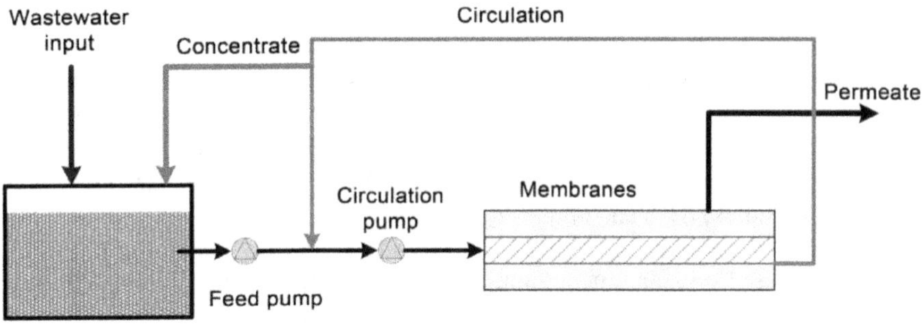

Figure 1.4. *Configuration in a separate tank with an external loop*

1.3. Types of membranes

Many suppliers on the market propose a wide range of products (Table 1.3). These membranes can be organic or inorganic. Organic membranes are generally made from hydrophobic polymer materials (polytetrafluoroethylene (PTFE), polypropylene (PP), etc.) or from hydrophilic polymer materials (polyethersulfone

(PES), polysulfone (PS), polyacrilonitrile (PAN), difluoride polyvinylidene (PVdF)). Membranes are therefore made by depositing material on a backing layer, and many variants exist. Manufacturers are currently offering various membrane structures, various membrane porosities and various types of materials.

The MBR market is dominated by the following membrane suppliers:

– flat sheet membranes: Kubota, A3, Toray, Martin System, Microdyn-Nadir, Huber;

– hollow fiber membranes: KMS-Puron, Zenon-GE, Mitsubishi Rayon, Memcor, Asahi Kasei, Polymem, Dow Chemical, Siemens, LG-Samsung, Tianjin Motimo;

– external loop membranes: Rhodia-Orelis, Norit X-Flow, Wehrle Environmental.

In France, large groups offer turnkey systems: Biosep® from Veolia, Aqua-RM® and Compact-RM® from Stereau and Ultrafor™ from Degrémont.

Membranes often consist of hollow fibers shaped as narrow porous tubes. Porous fibers are either grouped in a cylinder, thus making a tubular module, or in a bundle (hollow fiber module). In the tubular module, water to be treated enters at one end of the tubes and excess polluted water exits at the other end, while filtered water is recovered on the sides. The active part of the membrane is inside these tubes, and the water to be treated flows inside.

Two types of possible configurations of hollow fiber modules can be identified. In the first case, water flows inside the fibers and permeate is recovered outside the fibers (internal–external configuration). In the second case, water flows outside the fibers and permeate is recovered inside the fibers (external–internal configuration). Organic membranes can also be either stacked in layers separated by intermediate structures that ensure water circulation (flat modules), or self-rolled around a porous tube (spiral modules) (Berland and Juery 2002).

Among these organic membranes, there are membranes with symmetric, asymmetric and composite structures. Asymmetric membranes are composed of a thin layer known as "skin" deposited on a coarser porous backing of similar chemical nature, and whose role is to confer proper mechanical properties to the entire structure. Composite membranes differ from the previous ones as they are obtained by depositing the selective "skin" on a preexisting backing layer of different chemical nature. Symmetric membranes have a regular and uniform distribution of pores throughout their thickness (used only in microfiltration).

The materials used in membrane manufacturing must be resistant to the chemical products used for cleaning membranes, such as chlorine and certain acids or bases. Polypropylene membranes are sensitive to chlorine and must therefore be avoided.

There are advantages and disadvantages for each material used, as shown in Table 1.2.

Material	Advantages	Disadvantages
Polyvinylidene fluoride (PVdF)	Excellent flexibility Very high resistance Reasonable permeability	Difficult to modify its properties
Polyethylsulfone (PES)	Good chemical and thermal stability Easy to manufacture	Limited low pressures Low chemical resistance to aromatic compounds
Polyethylene (PE)	Inexpensive Good flexibility	Sensitive to oxidizing reactions
Polypropylene (PP)	Resistant to high temperatures	Moderate resistance to chemical agents Sensitive to chlorine

Table 1.2. *The main materials used for membrane manufacturing*

The main types of membranes used are microfiltration (MF) and ultrafiltration (UF). The nominal porosity for this type of application ranges between 0.01 and 1 μm.

Membranes can be layered or tubular. Filtration systems can operate under positive or negative pressure, depending on the manufacturer and on the intended application. For household wastewater applications, an immersed membrane filtration works under negative pressure in the biological tank. The membrane system works under positive pressure when it is implemented outside the biological tank. Microfiltration membranes are generally used for biological treatment of urban wastewater. Some manufacturers, however, propose ultrafiltration membranes that can remove viruses.

	Zenon	Puron	Mitsubishi	Kubota	Toray	Alfa Laval
Membrane type	HF	HF	HF	FS	FS	FS
Material	PVDF	PVDF	PVDF	PE + Cl	PVDF	PVDF
Pore size (μm)	0.04	0.03	0.4	0.4	0.08	0.2

Table 1.3. *Characteristics of membranes for six suppliers (HF = hollow fibers, FS = flat sheet membranes)*

According to its experience, each supplier has its own design recommendations (flux, membrane aeration, clogging control, chemical cleaning) and installation, implementation and operation procedures.

1.4. Description of general configurations of MBRs

Two large categories of membranes are present on the market:

– immersed membranes: filtration is achieved by imposing a transmembrane pressure (TMP) (between 0.1 and 1 bar) from outside the membrane towards the inside. Filtration is achieved by aspiration of the permeate, and membranes operate under vacuum;

– external membranes: activated sludge is pumped and recirculated at a tangential velocity (> 2 m/s) in the circulation loop. Filtration is tangential and is conducted by imposing a TMP (between 0.5 and 4 bars), from inside the membrane to the outside. The extracted permeate is evacuated to the exit/output channel and/or stored for cleaning membranes.

1.4.1. *Immersed membranes*

Immersed MBRs are a more recent configuration of MBRs with an external loop, aimed at reducing operation costs. The principle of this configuration consists of immersing the membranes in the bioreactor (activated sludge). The modules used are generally hollow fibers or flat sheet membranes. (Pseudo-frontal) filtration is achieved by applying negative pressure on the side of the permeate and by directly injecting air under the membrane module and inside the membrane module in order to maintain the particles in suspension and clean the external surface of the membrane. Given their low operating fluxes (15–30 $L.h^{-1}.m^{-2}$ compared to 50–120 $L.h^{-1}.m^{-2}$ for the MBR with external loop), several membranes must be simultaneously used in order to have a high membrane surface, which requires higher installation costs. Immersed membranes significantly reduce operating costs. This configuration has advantages in technical-economic terms where air injection has lower costs than a circulation of the suspension under turbulent conditions for the membranes with external loop. For a good transfer of oxygen, the MBR requires small bubbles of oxygen, while the membrane requires larger bubbles to maintain good permeability and reduce clogging phenomena. The appearance of immersed MBRs opened the municipal market of water treatment by membrane techniques. Immersed membranes therefore require higher installation capital, but they bring a significant reduction in operating costs compared to the external loop MBR. In order to improve the operating conditions, a third configuration is also used: the externally immersed MBR. In this configuration, the membrane is immersed in a tank outside the bioreactor. This has the advantage of maximizing the shearing stress in the vicinity of the membrane by the circulation of the sludge, on the one hand, and by the aeration of membranes, on the other hand. Consequently, clogging phenomena are reduced. This configuration allows decoupling the aeration dedicated to shearing from that dedicated to biological treatment and also facilitates the chemical cleaning phases. However, while the bioreactor requires good transfer of oxygen (requires small bubbles of oxygen), the membrane requires bubbles of larger dimensions, in order to maintain good permeability and reduce its clogging. For maintenance reasons, most manufacturers are now immersing membranes in a tank dedicated to filtration, which imposes significant recirculations (300–400% of the flow rate of permeate) in order to limit the increase in concentration in the membrane tank. The presence of the filtration tank increases investment costs (construction of the filtration tank) and operation costs (recirculation pump, additional aeration of the membrane in the filtration tank) and may call into question one of the key factors supporting the choice of MBRs at municipal scale.

The following categories can be identified:

– membranes immersed in a single tank: membranes are directly immersed in the biological tank;

– membranes immersed in a separate tank: the membranes are installed in another tank referred to as "membrane tank".

Membranes may consist of hollow fibers or flat surfaces.

Table 1.4 presents the advantages of both configurations. The separate tank configuration is generally preferred as it allows better control of the denitrification phase and facilitates the implementation of automated membrane cleaning operations. Indeed, the configurations in a single tank may yield poor denitrification performances because of the continuous aeration of membranes. Nevertheless, in the absence of strict requirements referring to nitrogen, and also for small plants, a single tank configuration reduces investment costs.

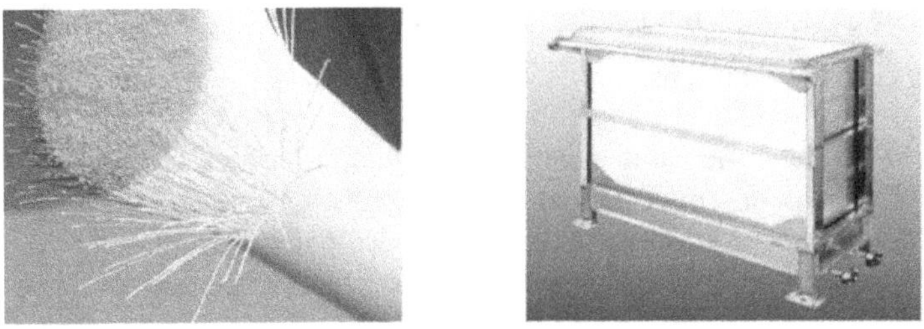

a) *b)*

Figure 1.5. *a) Hollow fibers. b) Flat sheet membranes (Toray)*

Single tank	Separate tank
No pumping required for recirculating the mixed liquor or feeding the membrane tank	Better control of the denitrification phase
Less costly implementation	Possibility to drain the tanks for exceptional cleaning (in terms of process and operation, prefer several small membrane tanks rather than a small number of large membrane tanks, in order to fully drain each tank and use the separate tank as a cleaning tank, while reducing the downtime of the station caused by these cleaning interventions)
For hollow fiber membranes, a separate immersion tank is needed to clean the membranes	The MES in the membrane system and in biology can be adjusted independently
Manual handling of hollow fiber membrane cassettes required for exceptional cleaning	The exceptional cleaning procedure can be readily automated
	No risk of hydraulic short circuit that may cause a biological clogging

Table 1.4. *Advantages of each configuration*

Certain systems with flat sheet membranes require a minimum MES concentration of about 12 $g.L^{-1}$ on the membranes. In the case of a separate membrane tank, the MES in the biological tank can be reduced to 5 $g.L^{-1}$ (e.g. off season, in a situation of temporary low load) while the membrane system can still operate with an MES concentration of 12 $g.L^{-1}$.

All the parameters involved in the design and operation of the MBR process influence membrane fouling. Three categories of factors are defined (see Figure 1.6):

– the characteristics of membranes and modules;

– the supply/feeding and biomass parameters;

– the operating conditions.

For some of them, the parameters have a direct influence on the fouling of MBR. Others lead to further effects on phenomena contributing to the development of clogging. The complex interactions between these parameters impact the perception and comprehension of clogging, and it is therefore important to have a proper understanding of the phenomena unfolding in MBRs in order to master the operation of this process and the means to implement in order to stop or reduce it.

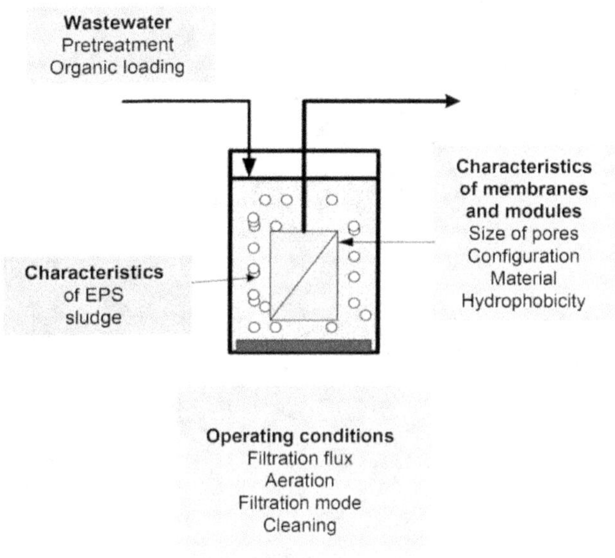

Figure 1.6. *Parameters affecting membrane clogging*

1.4.1.1. *Separate membrane tank*

In this configuration, the membrane tank is fed by the mixed liquor from the activated sludge tank. Figure 1.7 shows two cases of possible implantation of the membrane tank depending on the elevation of the biological tank.

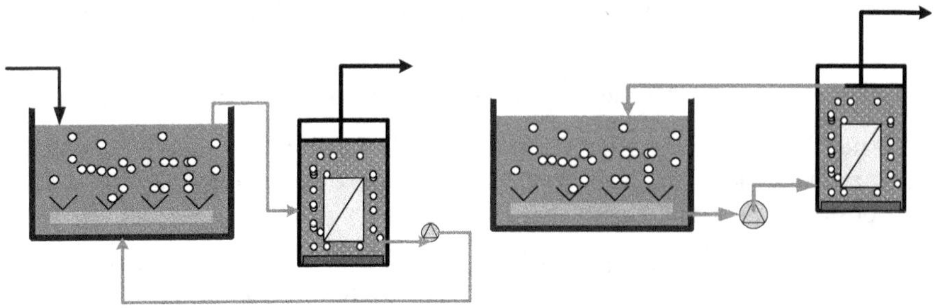

Figure 1.7. *Implementations of the configuration in a separate tank*

Figure 1.8. *MBR separate tank*

In case no. 1, there is gravity feeding of the membrane tank and pumped return. This configuration is often found in industrial units where the bioreactor is, for example, a 6 m high tank installed at height and the membrane system is enclosed in containers of about 3 m height.

In case no. 2, there is a pump feeding the membrane tank and a gravity return. This is the solution of choice for municipal stations, which is explained by the following reasons:

– The flux to the membrane system is easier (less costly) to control.

– In the configurations with several rows, all the filtration units are controlled exactly at the same water level (thanks to the overflow), which allows using the same boosters to aerate all membranes.

– In the case of foam formation in the membrane unit (due to intensive aeration), the foam may overflow to the bioreactor where it will be easier to manage than in a narrow membrane tank.

– This configuration makes it possible to establish a buffer capacity in the biological tank, which may be advantageous to control the peak flow.

In practice, since the permeate is only extracted from the membrane tank, it is necessary to recirculate the mixed liquor three to five times faster than the net filtration flow rate in order to stabilize the MES concentration to a reasonable value in the membrane tank.

The energy consumption of the recirculation pump may be significant. It is therefore important to minimize the water level difference between the bioreactor and the membrane unit(s), especially in the presence of high flow rates.

Activated sludge is sent to the membrane tank through pipes installed at the base of the tank and having holes on their lower or lateral side, in order to avoid damaging the membranes. This piping setup allows a better distribution of the mixed liquor exiting the biological tank.

It is interesting to place the membranes in compartments specific to each row of cassettes or modules. If the membrane tank is crossed by a significant recirculation flow (typically five times the flow rate of the station), there is a risk of “leaching” the chemical products when cleaning operations take place, and therefore limiting the efficiency of maintenance cleaning. A membrane tank whose rows are not separated by waterproof linings does not allow advanced cleaning in place without using an external tank.

1.4.1.2. *Single tank*

In this configuration, membranes are directly immersed in the biological tank (see Figure 1.9). This is the simplest configuration as it does not require a membrane tank or a recirculation circuit.

Exceptional cleaning is conducted in a dedicated cleaning tank placed outside the biological tank with modules of the same row aligned; it is therefore necessary to have a means of lifting modules to place them in this tank, either a mobile one (e.g. a lifting crane) or a fixed one (pillar crane, portal, etc.).

To avoid compromising the effectiveness of maintenance cleaning (with the use of chemical products), membranes should not be placed on top of or too close to air process diffusers in the bio tank.

Indeed, the airlift and the spiral flow induced by the diffusers can “wash” the chemical products, and limit their time of contact with the outside of membranes. If this is the case, it is imperative to be able to stop the air process under the membranes during cleaning.

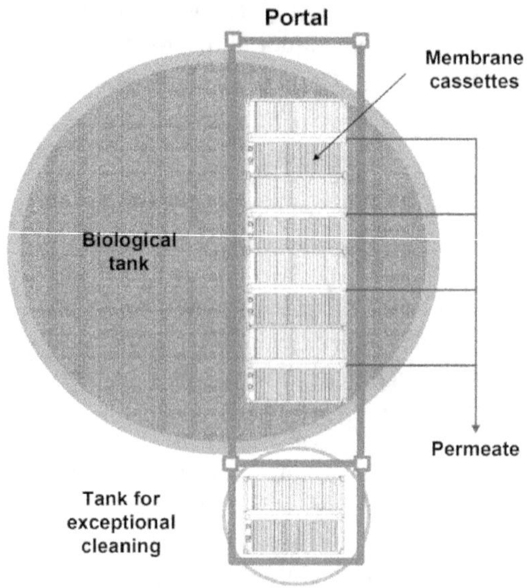

Figure 1.9. *Example of implementation of a single tank configuration with a tank reserved for potential exceptional cleaning*

Permeate can be extracted by pumping, gravity or siphon (see Figure 1.10).

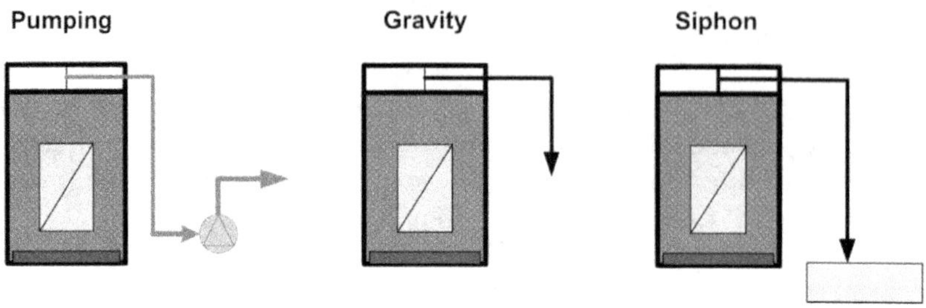

Figure 1.10. *Three configurations for the extraction of permeate*

1.4.2. *MBR with external loop*

The MBR with an external loop is the first generation of MBRs with tubular membranes implemented in the external loops. The biomass circulates at high speed in these loops in order to limit membrane clogging and reduce the frequency of chemical cleaning. The biomass retained by membrane filtration is recycled in order to maintain a much higher concentration in the bioreactor. The MBR with an external loop is generally used for the treatment of wastewater characterized by relatively high temperatures (e.g. 40°C), high organic loads (e.g. 10,500 mg DCO.L^{-1}) or high pH.

Several treatment stations are equipped with (generally tubular) membranes assembled on structures installed outside biological tanks (these membrane systems are referred to as "with external loop") for fluxes > 1,000 $m^3.d^{-1}$ as the operating costs of these units (energy consumption for the membrane operation) are much higher than in the presence of a configuration in which membranes are directly immersed (energy consumption for the operation of the membrane typically < 0.5 kWh.m^{-3} of permeate). Their disadvantages are essentially related to the significant rates of shear undergone by the biomass and also the relatively high energy consumption (of about 10–50 kWh.m^{-3}). The MBR with an external loop generates significant operating costs. However, these costs can be justified under specific conditions, such as the treatment of wastewater characterized by very high loads and the implementation of intensive systems.

Two configurations are possible and presented in Table 1.5. The systems withn external loop are relatively simple to implement and are less sensitive to clogging in the case of limited filterability of the biomass. Constructors often justify their choice by explaining that membranes combine a small size of pores with high permeability to provide a high quality effluent at low operational cost.

The main characteristics of membranes associated with this type of MBR are:

– PVDF(polyvinylidene difluoride)/PES (polyethylsulfone) membranes;

– small size of pores;

– high permeability;

– pressurized and closed systems;

– robustness;

– small dimensions;

– pore size of 30 nm;

– high quality instantaneous effluent;

– MS2 bacteriophage 4-Log tested.

Inserted in a protective tube, the membrane is optimally protected and does not risk abrasive wear of fibers during operation. The separation layer is entirely isolated from external influences.

	Crossflow	*Air lift*
Diagram	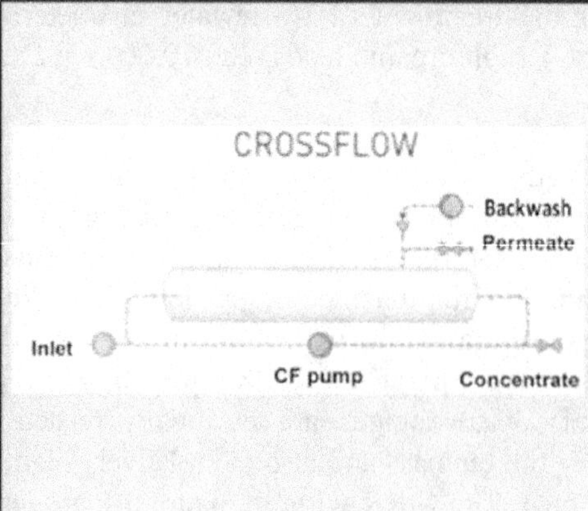	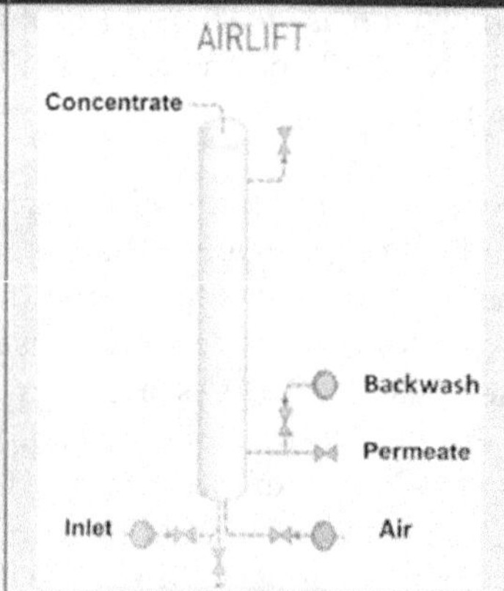
Specifications	MES: 12–30 g.L^{-1} Flux: 80–200 L.m^{-2}.h^{-1} Reduced overall size Energy consumption: 1.5–4 kwh.m^{-3} Continuous operation PTM: 1–5 bar Membrane: 1.5 mm diameter Length: 3 m	MES: 8–12 g.L^{-1} Flux: 45–65 L.m^{-2}.h^{-1} Reduced overall size Energy consumption: 0.25 kwh.m^{-3} Discontinuous operation PTM: 0.05–0.3 bar

Table 1.5. *Characteristics of MBR with external loop*

MBR membranes are distinguished by their structure and their manufacturing process. Fiber protection differs depending on whether they are potted in tubes or in free air. It may consist of a thin protective layer relative to the solid materials that may degrade the external surface, or the membranes may be without protection when they are potted in tubes.

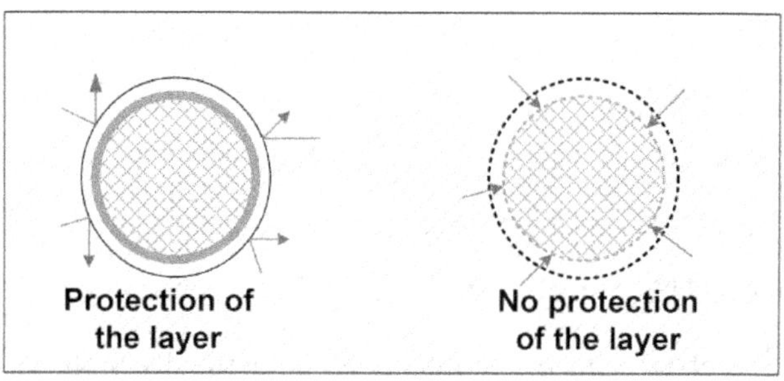

Figure 1.11. *Protection of fibers in tubular membranes*

Systems with tubular membranes can be used for:

– easy drainage of the system to remove the retained particles and sludge;

– safe, effective and efficient chemical cleaning;

– easy handling.

a) b)

Figure 1.12. *X-flow tubular membranes: a) vertical; b) horizontal*

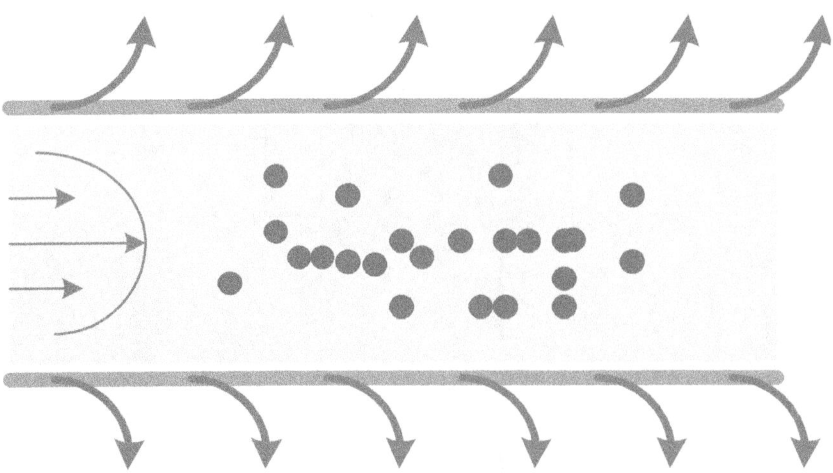

Figure 1.13. *X-flow tubular membranes: filtration direction*

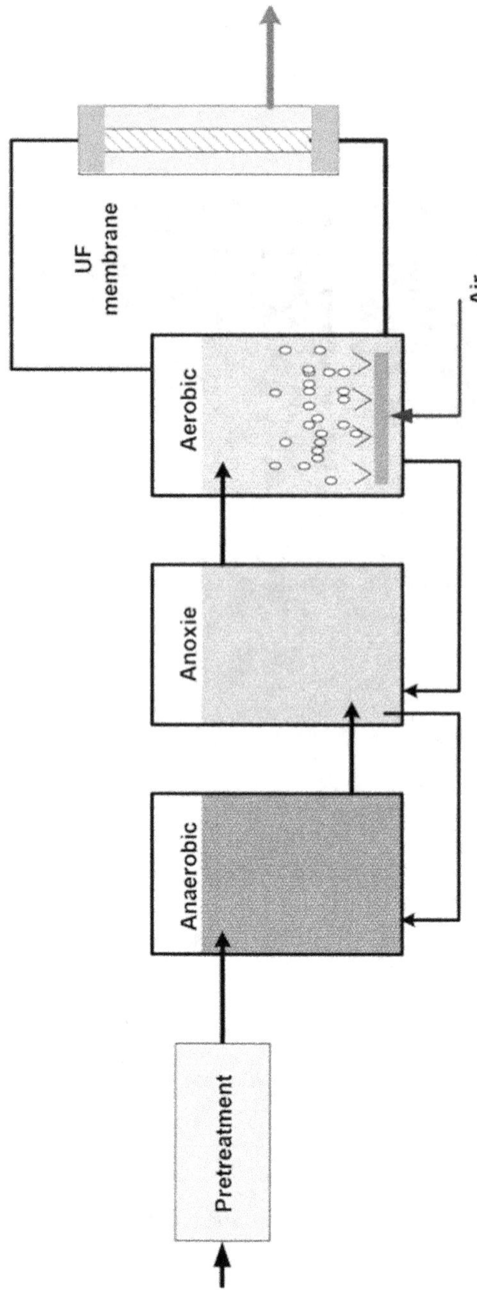

Figure 1.14. *Diagram of a biological treatment system with external membranes*

These membranes operate based on a cross flow inside the tubes. Filtration in the fibers has an internal–external direction. A continuous airlift avoids the accumulation of solids on the membranes and allows the recycling of sludge in the biological tank.

In a system with an external loop (see Figure 1.14), the acceptable concentration of sludge in the biological tank ranges between 10 and 20 g.L^{-1}. The quality of the output effluent gives an MES concentration < 3 mg.L^{-1}.

The operating flux is 50 lmh, PTM varies between 0.1 and 0.3 bar and permeability is 500 lmh.bar. The term lmh is an abbreviation of L.m^{-2}.h^{-1}.

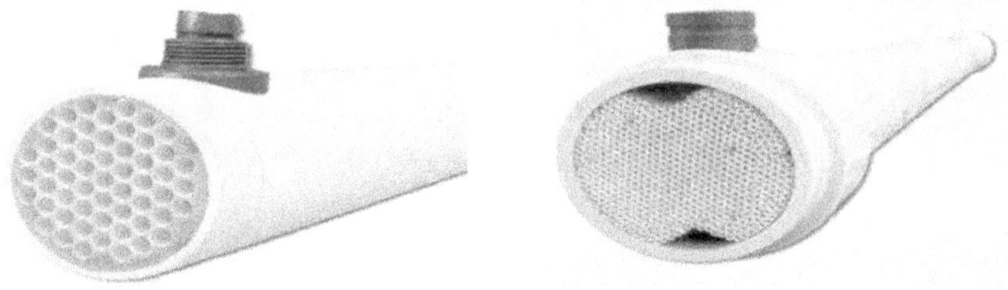

a) *b)*

Figure 1.15. *Examples of membrane modules operating: a) in cross flow; b) helped by an airlift*

1.5. MBR dimensioning parameters

1.5.1. *Mechanical pretreatments*

Membrane filtration requires advanced pretreatment of raw waters in order to limit the obstruction of physical clogging of membranes. The recommended pretreatment involves the following stages:

– coarse screening (10–25 mm);

– grit removal (up to 200 μm);

– very fine screening.

The screening stage is very important and has a direct consequence on the lifetime of membranes, as it allows retaining the fibers that are an aggravating factor for the risk of membrane solidifying (solidifying being an accumulation of very dense sludge between the fibers leading to a significant loss of filtration surface).

The hollow fiber membranes are more sensitive to the accumulation of fibers than flat sheet membranes and therefore require a finer mesh opening.

In urban wastewater, we can systematize the use of rotary screens, with round meshes of (maximum) 1 mm for hollow fiber membranes, while round meshes ≤ 3 mm must be sufficient for flat sheet membranes.

Figure 1.16. *Rotary screens (1 mm)*

1.5.2. *Buffer tank*

a) *b)*

Figure 1.17. *Clogging generated by a) poor screening; b) good screening*

The implementation of a buffer tank is a good solution for reducing the filtration surface required for the treatment of incoming effluents. The buffer tank is dimensioned to store, during the peak hours, the volumes of raw water that the MBR cannot take in. Its volume is calculated by multiplying the difference between the peak flow rate and the average flow rate by the total duration of these peak periods:

$$\text{Buffer tank volume} = (Q_{\text{peak flux}} - Q_{\text{net}}) \times \text{peak duration}$$

For example, if the daily average maximum flow rate is 1,440 $m^3.d^{-1}$, the flow rate to be taken into account for the calculation of the filtration surface will be 1,440/24 or 60 $m^3.h^{-1}$. The net permeate flux is therefore 60 $m^3.h^{-1}$. If the peak flow rate is equal to 140 $m^3.h^{-1}$ and the peak duration to hours, the volume of the tank is calculated as follows (see Figure 1.18):

$$\text{Buffer tank volume} = (140 - 60) \times 3 = 240 \text{ m}^3$$

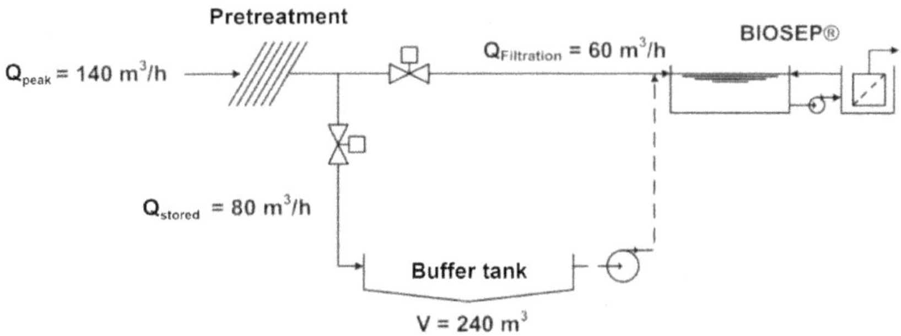

Figure 1.18. *Example of calculation of the buffer tank volume*

1.5.3. *Other factors to be taken into account*

1.5.3.1. *Silicone*

Silicone is a highly clogging product. It is encountered in effluents resulting from textile or cosmetic industries. It is not degraded by biology and accumulates in the tank, depositing on all equipment and primarily on membranes. Filtration then proves to be very difficult with a rapid drift of filtration performances (down to losing the treatment capacity).

On this type of effluent, it is not conceivable to propose an MBR without a physico-chemical pretreatment (coagulation–flocculation + decantation or flotation).

The dimensioning of pretreatment, and particularly the determination of the reagent dosage, can only be done based on tests (Jar Test/Flotation test).

1.5.3.2. *Grease/oils/hydrocarbons*

When the grease or oil content exceeds a value of 100 mg.L^{-1} before pretreatment, there should be a flotation stage upstream of the MBR.

In the presence of hydrocarbons, a pretreatment by adsorption by flotation is required to reach a target value of 20 ppm at the inlet of the biological tank.

1.5.3.3. *Hardness*

Hard effluents are problematic because of the chlorine-based maintenance cleaning. These are conducted with a solution of sodium hypochlorite. Ca^{2+} and CO_3^{2-} ions precipitate. If alkalinity ($CaCO_3$) is high, the formation of these precipitates is all the more significant and is reflected by a loss of filtration of membranes and their calcification. This precipitation phenomenon can occur even at low alkalinities (alkalinity < 200 mg.L^{-1} $CaCO_3$), but frequent maintenance cleanings with acid can stabilize filtration performances.

On the other hand, for higher values (TH > 500 mg.L^{-1} $CaCO_3$), the maintenance cleanings with acid are not sufficient: raw water pretreatment by softening should be conducted.

1.6. Tank–membrane association

1.6.1. *Parameters linked to the biological tank*

MBR implementation requires taking into account a number of factors that govern each basic operation, particularly MBR dimensioning, hydrodynamic operating conditions, membrane structure, membrane module configuration, membrane cost, etc. The hydraulic retention time (HRT) will determine the time available for degrading soluble compounds, and therefore the volume of the bioreactor. The solid retention time (SRT, or sludge age: θ) corresponds to the time required to degrade particulate compounds and allow renewal/lysis of the purifying biomass. It will therefore directly influence the MES content of the tank.

MBR is a discontinuous treatment process whose operation is automated. It alternates:

– sequences of permeate production (filtration);

– sequences that do not produce and even consume permeate (backwashing, relaxation, degasification, chemical cleaning, etc.).

The sequences required for the proper operation of membranes are generally defined by the suppliers. Backwashing, relaxation, degasification and cleaning sequences are aimed to maintain or recover the filtration performances. The phases of backwashing, cleaning and possibly of degasification, depending on the technique used, consume permeate. When the solid–liquid separation is conducted by membrane filtration, all materials in suspension are retained by the membranes. In this case, the decantation capacity of the sludge has no impact on the efficiency of the solid–liquid separation. Unlike decanters, biopolymers can induce clogging of membranes, reduce the filtration capacity or have an impact on the frequency of cleanings. However, to the extent that sludge retention time is sufficient to induce good nitrification, biopolymers do not seem to cause significant clogging. Hence, in the case of MBR, the age of sludge can be maintained between 20 and 70 days and the MES concentration can reach 20,000 $mg.L^{-1}$, depending on the documentation. However, at concentrations of this order, sludge density limits the filtration rate of membranes and also the oxygen transfer due to viscosity phenomena. For these reasons, the MES concentration in MBRs is limited to 10,000 or 12,000 $mg.L^{-1}$ (Atasi 2006).

Since the capacity of sludge to decant is not an important factor during the design or operation of this type of treatment system, the factors that have an influence on sludge settling are not considered. This type of system is far simpler to operate. The intensity of stirring required in the biological reactor increases with the concentration of MES that must be maintained in suspension. Hence, in an MBR, the installed stirring capacity must be sufficient to maintain in suspension over 12,000 $mg.L^{-1}$. When denitrification precedes the treatment (in order to save aeration energy), the volume of the pre-anoxic zone must be increased in order to compensate the supplies of oxygen by the sludge recycling system.

Sludge properties and their impact on membrane fouling are explained through the particle size distribution and the major constituents considered as being involved in the fouling process, namely the extracellular polymer substances and the soluble products resulting from microbial activities.

The constituents of activated sludge are generally classified into three groups: materials in suspension (> 1 μm); colloids (0.1–1 μm); soluble constituents (< 0.1 μm).

The size of the physical characteristics of activated sludge has a major role in the fouling process. However, it is currently admitted that the size of flocs of activated sludge has no direct correlation with fouling. This is mainly due to the fact that the turbulence promoted in the proximity of the membrane surface by the aeration prevents large deposits of particles. Colloids are considered today as the main contributor to fouling given their size (0.1 and 0.5 μm), near the size of pores of the membrane.

Extracellular polymeric substances (EPSs) provide the cement that keeps together the microbial aggregates. They form a layer that protects microorganisms from external influences.

The biology–membrane association requires parameters identical to those of a classical activated sludge for the biological tank part.

1.6.1.1. *Hydraulics*

Hydraulic parameters to be considered are the daily average flow rate during the rainy season (unit network) and the daily average flow rate during the dry season (separate system). The peak hour coefficient is also taken into consideration.

This is why the biological tank is often preceded by a tank that stores the peak hour surplus, which allows the optimization of the membrane surface.

1.6.1.2. *Nature of pollution: characteristics of water to be treated*

1.6.1.2.1. Activated sludge tank

Hydraulic parameters to be considered are the volume of the biological tank, concentration of oxygen dissolved in the biological tank, rate of recirculation of the mixed liquor and also the age of sludge.

The biological reactor is modeled as a classical tank of activated sludge except that the higher mixed liquor suspended solid (MLSS) concentration (< 12 $g.L^{-1}$) allows the reduction of dimensions of biological tanks. The shape of the tank (single tank, anoxic + aerobic tanks, channel) is independent of the membrane filtration.

In the case of a "separate tank" configuration, the activated sludge is recirculated between the biological tank and the membrane tank, thus forming an additional tank that should be considered as an aerated volume (oxygen concentration about 4 $mg.L^{-1}$). The system will be chosen depending on the efficiency and quality of discharge.

1.6.1.2.2. Non-constraining level of total nitrogen (TN) (15 $mg.L^{-1}$) and 75% efficiency

A complete mixing solution is sufficient. Aeration is controlled by a redox probe or directly by NH_4^+/NO_3^- concentrations.

Figure 1.19. *Clogging of immersed membranes*

1.6.1.2.3. Constraining level of TN (10 mg.L^{-1}) and efficiency above 85%

Applying these operating conditions requires a solution in two (anoxic and aerobic) tanks to allow better control of denitrification. The process using an oxidation channel is also possible by controlling aeration and anoxic periods.

Figure 1.20 shows an example of the MBR configuration.

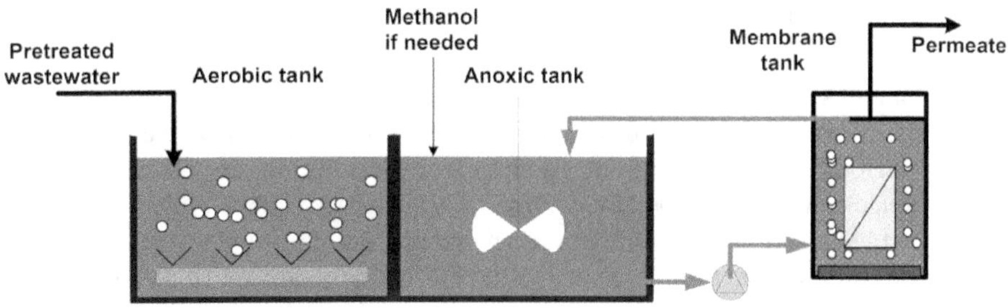

Figure 1.20. *Aerobic–anoxic configuration of MBR*

The membrane system does not instantly adjust to the variations of input flux/flow (there is no overflow as in the classical activated sludge plants). At least

15 minutes are required to make the permeate extraction system pass from the minimum flux to the maximum flux. The biological tank should therefore have the available sufficient buffer capacity. In general, a 25 cm level difference in the bioreactor is appropriate.

1.6.1.3. *Hydraulic retention time (HRT)*

This parameter determines the time available to degrade soluble compounds, and therefore the volume of the bioreactor. Besides the abovementioned factors, the choice of characteristics of the filtration system must also integrate the specific nature of suspensions present in the MBR, and particularly their significant concentration in cell biomass (10–15 $g.L^{-1}$), as well as their evolving nature depending on the imposed hydrodynamic and biological constraints. Indeed, the biomass concentration in MBRs is not limited to a critical value close to 4–5 $g.L^{-1}$, beyond which the separation by gravity decantation/settling appears as strongly slowed down, even perturbed, in the conventional systems with activated sludge. MBRs are thus susceptible to operate with concentrations that are two to five times higher (between 8 and 25 $g.L^{-1}$) depending on the configuration.

1.6.1.4. *Sludge age*

Similar to other biological processes, the age of sludge in MBRs determines the quality of treatment, but it also influences the filtration parameters. It is thus important to stick to a minimum sludge age (independent of temperature): it is the solid retention time (SRT, or sludge age) that corresponds to the time required to degrade the particulate components and allow the renewal/lysis of the purifying biomass. It will therefore directly influence the MLSS content of the tank. Furthermore, a proper biological operation requires making sure there is no limiting oxygen.

– Minimum sludge age: 15 days. This parameter must allow the complete oxidation of organic matter, in order to avoid the rapid clogging of the membrane. This is verified by measuring the ammonia concentration at the inlet of the membrane tank, which must be below 2 $mg.L^{-1}$. In practice, this is typically reflected by a sludge age above 15 days. Moreover, an operation at high sludge age (> 15 days) offers better filtration performances, while too low a sludge age (< 15 days) generally produces a heavily clogging sludge.

– Maximum sludge age: 80 days. Beyond this sludge age and according to the influent load variations, there will be an increase in the risk of appearance of extracellular polymer substances (EPSs) that contribute to membrane clogging.

Figure 1.21 shows the evolution of the correction factor allowing the calculation of the sludge age equivalent to various temperatures based on the sludge age at 12°C:

$$G(T\ °C)/G(12°C) = 2.3333\ e^{-0.0705 \times T}$$

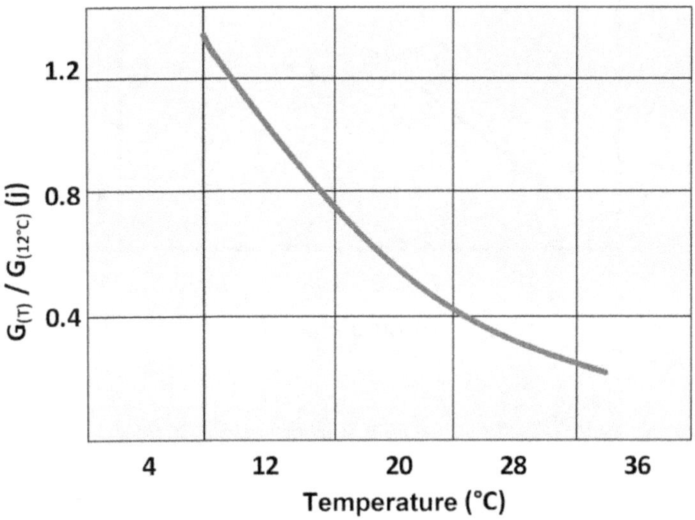

Figure 1.21. *Sludge age correction factor depending on temperature*

Stirring is required for:

– ensuring sludge homogenization;

– avoiding sludge decantation;

– limiting the risks of sludge overconcentration at the level of modules.

The stirring equipment depends on the type of equipment chosen for the oxygen transfer function in the process air supply.

The recirculation loop permanently ensures the supply of the membrane tank with mixed liquor from the biological tank. This recirculation allows the minimization of sludge overconcentration occurring in the membrane tank during filtration. The calculation of the concentration in the membrane tank and in the biological tanks relies on a simple principle: membranes remove interstitial water between SS and therefore the MLSS concentration in the mixed liquor. A mass balance on the membrane tank makes it possible to find the various input/output concentrations depending on the recirculation rate.

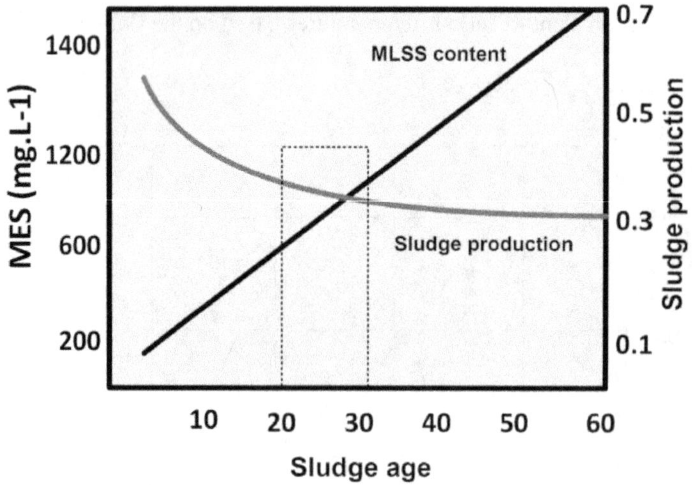

Figure 1.22. *Influence of operating parameters on a membrane bioreactor operation (Lobos 2009)*

The typical recirculation flow rates of the mixed liquor at the inlet of the membrane tank are four to five times the net flow rate. A high recirculation rate must be maintained between the biological reactor and the filtration tank. A 200–400% recirculation is recommended to prevent a significant increase in the MLSS concentration in the filtration tank, which might cause a decrease in the filtration rate and the system failure. When membranes are directly placed in the biological reactor, recirculation is not required, but a tank for cleaning membranes is required.

EXAMPLE.–

The daily flow rate (daily Q) is 384 $m^3.d^{-1}$. As the recirculation flow rate is dimensioned on the net flow rate to be treated (Qnet), the recirculation flow rate is calculated as follows:

$$\text{Qnet to be treated} = \text{Q daily}/24 = 16\ m^3.h^{-1}$$

For domestic wastewater, we consider five times the net flow rate to be treated:

$$\text{Q recirculation} = \text{Qnet to be treated} \times 5 = 16 \times 5 = 80\ m^3.h^{-1}$$

The choice of recirculation pumps is based on the following aspects:

– Recirculation pumps may represent a significant part of the total energy consumption of the plant. Therefore, good efficiency for the pump presents a specific interest in this case. If the water level in the aeration tank is close to the one in the membrane tank, propeller pumps can be used.

– The mixed liquor to be pumped (particularly during the transfer of the filtration tank to the biological tank) may contain a significant quantity of air. It is essential that recirculation pumps tolerate these conditions.

During exceptional cleanings, the sludge activated in the membrane tank must be drained and returned in the biological tank or in the adjacent membrane tank, or sent to the sludge treatment. As for the chlorine solution drainage, it can be moved under gravity to the all-water station by means of a manual valve or by pumping using a special mesh with the pump backflow. The acid solution is drained similarly to the chlorine solution except that the solution is neutralized with soda before discharge. The overflow of the mixed liquor of the biological tank to the membrane tank must be closed during cleaning to allow this same tank to store the excess mixed liquor.

It is essential to verify that no residual BOD ever remains in contact with membranes to avoid biological clogging at the surface of membranes.

In a single-tank immersed configuration, the risk of hydraulic bypass of one part of the effluent, or of solutions with methanol addition for denitrification, must be studied very carefully to ensure the quality of the treated water.

In a separate-tank immersed configuration, the fraction of hard COD should not exceed 2–3 $mg.L^{-1}$ in the membrane tank (total soluble COD < 20–40 $mg.L^{-1}$). The addition of an aerated tank upstream of the membrane tank can be considered to obtain these values.

Because of a higher sludge age, the sludge production of MBRs is theoretically lower than that of classical activated sludge, but the physical barrier of membranes retains more SS, which increases the sludge production. The sludge production by an MBR process compared to that of a classical activated sludge process shows a 5–8% increase in the quantity of sludge to be extracted for the membrane process. This corresponds to a drop by 12–20 $mg.L^{-1}$ of total SS in the effluent of MBR with respect to the classical activated sludge. With a domestic effluent, sludge production is close to 0.45 kg SS/kg $COD_{removed}.d^{-1}$.

In an MBR, the relatively high content of suspended solids in the biological tank drives higher viscosity values compared to conventional systems. Activated sludge is considered a non-Newtonian fluid behaving as a pseudo-plastic fluid (Rosenberger et al. 2002; Le-Clech et al. 2006).

A model based on the Ostwald model is proposed to determine the apparent viscosity as a function of the MLSS content:

$$\eta = \exp(0.882.\ \text{MLSS}\ 0.494) - (\kappa) - 0.050.\ \text{MLSS}\ 0.631$$

with:

– η: apparent viscosity of activated sludge (Pa.s);

– MLSS: mixed liquor suspended solid concentration in the biological tank ($g.L^{-1}$);

– κ: shearing rate at the level of the membrane (s^{-1}).

Aeration is used in the immersed MBR configurations to provide oxygen to the biomass, to maintain activated sludge in suspension and to reduce the clogging by turbulences and the cleaning of the membrane surface, that is, avoid membrane clogging and limit as much as possible the appearance of fouling.

Optimal aeration improves the filtration conditions up to an ideal theoretical value without, however, leading to an improvement in the membrane treatment performances.

The actual oxygen needs are calculated in the same manner as for classical activated sludge. They are slightly higher because of the higher sludge age. The specificity of MBR comes from the strong MLSS concentration in the tank, which impacts the calculation of the alpha factor. The global factor is calculated by taking into account MLSS concentration and the temperature of the effluent. Aeration can thus become a limiting parameter for reducing the dimension of the biological tank. Measures should be taken to make sure that the biological tank is sufficiently large to receive the aeration system.

Figure 1.23 presents data recorded for one year of evolution of permeability observed for a municipal MBR. The problems inherent to aeration during the first months result in a clear drop in permeability and in the flow rate of filtered water.

The MBR process relies on two aeration systems: the air process that provides the oxygen needed by the active biomass to ensure a proper biological treatment, and the air membrane whose function is to avoid membrane clogging.

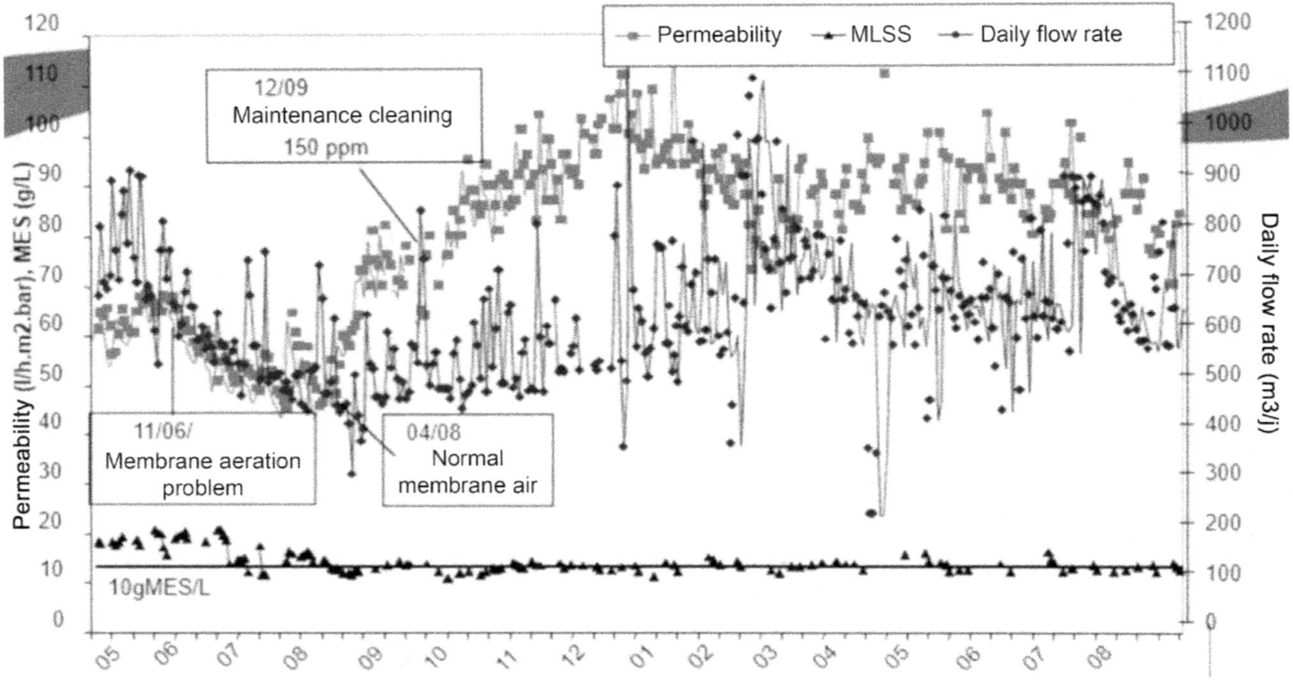

Figure 1.23. *Problems inherent to aeration. For a color version of this figure, seewww.iste.co.uk/gaid/wastewaterengineering2.zip*

Membrane aeration can be based on continuous or sequential operation according to recommendations of the membrane supplier. It generally involves large bubble diffusers, and more recently fine bubble diffusers. The operation of the membrane aeration is related to the performances of membranes and to the operation cycle, but it is independent of the biological process, which is susceptible to perturbing the efficiency of the treatment as the membrane air can, for example, introduce an additional quantity of oxygen in the biomass during denitrification.

In the separate-tank immersed configuration, the low volume of the membrane tank compared to the significant injected quantities of air leads to an oxygen concentration in the membrane tank from 2 to 4 $mg.L^{-1}$.

In the single-tank immersed configuration, the major disadvantage is the impossibility of reducing the oxygen concentration in the tank, and consequently the absence of actual ranges of denitrification. In a period of underload, membrane aeration may be sufficient to meet and even exceed the air process needs. When process aerators operate, the oxygenation capacity produced by the membrane air is generally considered negligible. However, during denitrification phases, when the oxygen concentration is low, the membrane air is no longer negligible.

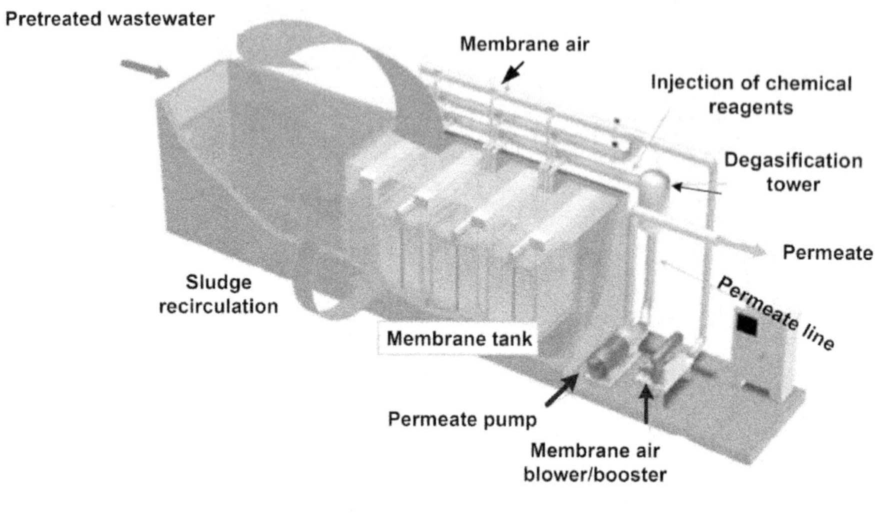

Figure 1.24. *Devices of the tank–membrane pair in a single tank*

Sludge can be extracted in two different places:

– Sludge can be extracted directly from a biological tank near an aeration zone using a volumetric pump. In fact, extraction must be conducted during an aeration phase to allow the proper homogeneity of extracted sludge. If a centrifugal pump is used for sludge extraction, we should make sure there are no risks of air pumping, either by placing a deflector at the level of the suction point during design, or by placing a suction point at more than one meter from the aeration zone.

– In the case of the immersed configuration with a separate tank, it is preferable to remove excess sludge at the outlet of the membrane tank. This way, it is more concentrated and properly aerated (therefore compatible also with the biological removal of phosphorus).

Extraction must be conducted as continuously as possible in order to minimize the potential abrupt changes of concentration in the biological tank. For this purpose, one continuous extraction of a small volume of sludge over a long period is recommended more than a rapid extraction of a significant volume only once a day.

Sludge coming from the biological tank can be sent directly to a centrifuge due to the high concentration (10 $g.L^{-1}$). It should be noted that sludge coming from MBR is generally difficult to decant, rendering densification more difficult.

1.6.2. *Parameters related to the membrane*

Due to the physical barrier formed by the membrane and the size of pores, a fraction of the soluble pollution (large size molecules) is retained by the membrane, while in the case of classical activated sludge, they exit the system with the treated water. Hence, this pollution, considered soluble, becomes particulate in the case of MBRs. The implementation of membranes requires taking into consideration a certain number of factors that govern this essential phase, particularly the hydrodynamic operating conditions, the structure of the membrane, the configuration of the membrane module, the cost of the membrane, etc.

The optimization of the MBR design and operation requires perfect knowledge of the relations between operating parameters and biological parameters (hydraulic retention time, sludge age, TMP, critical flux, efficiency, biomass activity, biomass concentration, etc.), as well as the physical parameters (filtration, durability, clogging of membranes, backwashing, etc.). The implementation of biological processes involved in the biodegradation of recalcitrant compounds requires an acclimation of the cell biomass to the degradation of pollution.

One of the major challenges of the industrial implementation of MBRs is maintaining during operation the membrane permeability at economically acceptable values, and therefore minimizing clogging. In general, the parameters monitored during operation are TMP, the water flow velocity (v), the concentrations of SS in the water to be treated and in the bioreactor, the hydraulic retention time (HRT), the concentration of dissolved oxygen, the specific consumption of oxygen, etc.

Ratios concerning the transfer of the soluble compartment to the particulate compartment for this application are used by the constructors in this fraction at the level of COD, BOD_5 and NK.

For the SS parameter, the concentration in conventional SS is increased to take into consideration the retention of dissolved materials (> 0.1 μm), which in a conventional activated sludge treatment undergo clarification.

The applied flows/flux must be carefully chosen to avoid the risk of rapidly driving a drop in permeability and too frequent cleaning, leading to a reduction of membrane lifetime. The velocity of the cross flow or of the transmembrane flow influences clogging and the nature of membrane clogging. When the MBR with an external loop operates at a low cross-flow velocity, between 1 and 2 $m.s^{-1}$ for an ultrafiltration membrane and between 1 and 3 $m.s^{-1}$ for a microfiltration membrane, the unfolding clogging is reversible. On the other hand, for velocities ≥ 3 $m.s^{-1}$, clogging becomes irreversible. A stable flow can be maintained during a long filtration period, when the MBR operates at a reasonably adjusted TMP.

Many studies indicate that the flow through the membrane or the cross flow and the TMP must be maintained below respective critical values. These critical values must be determined for each type of effluent, membrane and configuration of the membrane module (Cicek et al. 1998; Defrance 1999b).

Figure 1.25 shows the safe dimensioning zone for an MBR with immersed membranes with flows/flux depending on the peak frequency. The determination of the filtration surface is based on the flux/flow criterion (acceptable flow rate per m^2 of membranes) at a given temperature, the temperature affecting the filtered flow rate through the modification of water viscosity. The dimensioning fluxes considered for calculating the filtration surface are net fluxes. They must be chosen as a function of the:

– membrane supplier (each supplier has its own design recommendations);

– temperature of the mixed liquor;

– peak frequency (the higher the frequency, the more conservative the flux should be).

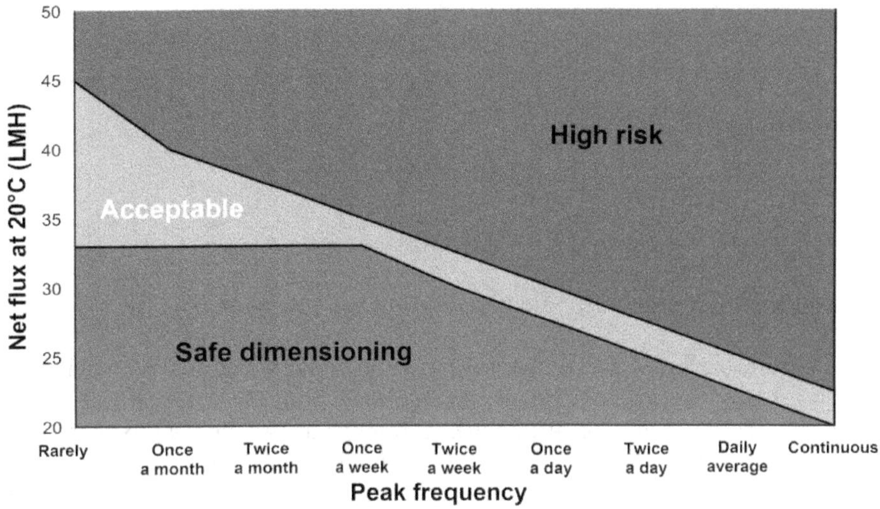

Figure 1.25. *Choice of fluxes applied to immersed membranes*

At 20°C and for standard projects, dimensioning fluxes may go from 10 to 40 L.m^{-2}.h^{-1}, according to the abovementioned conditions.

The filtration surface is calculated based on the limiting net flux. The hourly flow rates higher than the flow rate admitted on the membranes lead to one part of the influent being stored in a buffer structure upstream of the biological tank.

The hydraulic parameters used for the dimensioning of members are:

– the instantaneous filtration flow rate (Qi) or raw flow rate, expressed in m^3.h^{-1}, which denotes the instantaneous flow rate sucked by a row during the filtration phase. It is the set point of the flow rate applied to the pump or to the control valve;

– the net flow rate of the MBR system (set of rows/files, generally operating under the same conditions) which represents the average hourly flow rate actually treated by the membranes, which is the average flow rate discharged or available having subtracted all the permeate needs that are essential to the system operation (e.g. backwashing, cleaning, degasification). It is calculated by means of a balance of volumes on the tank permeate over 24 h;

– the net flow rate and the instantaneous flow rate, which are linked by the hydraulic efficiency, also representing the ratio between the quantity of water exiting the system per unit time and the quantity of filtered water per unit time. This corresponds to the production efficiency of the system:

$$Rh = \frac{Qnet}{Qi} = \frac{Fluxnet}{Flux\ i}$$

Qnet ($m^3.h^{-1}$) = Qj produced by the membranes – Qj consumed by the membranes

In general:

Qj net = Qj discharge/reuse + Q industrial water

A temperature correction should be applied to the flow rates similarly to the flux.

The permeate flux corresponds to the through flow per m^2 of membranes. Similarly to how the net flow rate is distinguished from the instantaneous or raw flow rate, the net flux (Fnet) is distinguished from the instantaneous flux in filtration (Fi) or to the raw flux. The value of a flux is significant only if brought at a given temperature, as the temperature of the fluid influences its filterability, through the modification of its viscosity. The common approach to comparing the hydraulic performances obtained at various temperatures is to normalize the operating flux at a reference temperature (often 20°C). This might be done by applying a temperature correction factor. To avoid the interference of temperature effects on MBR fouling, a nonlinear regression between the critical flux and temperature is proposed:

$$Fi,T = Fi,20 \times 1.025^{(T-20)}$$

$$Fnet(T) = \frac{Qnet(T) \text{ x } 1000}{Sf}$$

$$Fnet(20) = \frac{Fnet(T) \text{ x } \eta(T)}{\eta(20)}$$

$$Fi(T) = \frac{Qi(T) \text{ x } 1000}{Sf}$$

$$Fi(20) = \frac{Fi(T) \text{ x } \eta(T)}{\eta(20)}$$

In Europe, the flux is expressed in liters of permeate produced per m^2 of membrane and per hour (or $L.m^{-2}.h^{-1}$ or lmh). There is another international unit, the "m/j", which corresponds physically to the water approach velocity at the contact with membranes. It is therefore essential to always specify the temperature associated with the flux.

The design of the required filtration surface of the membrane systems is mainly based on the filtration rates. In an MBR, the net filtration flux can range between 14 and 25 $L.m^{-2}.h^{-1}$ if the concentration of MLSS in the reactor remains below 12,000 $mg.L^{-1}$ at 20°C. This flux must be reduced by 3% per °C below 20°C. Because of periodical backwashes, the net filtration flux is below the instantaneous flux according to a ratio that can vary between 1.1 and 1.2 depending on the programmed washing sequences.

As the membrane system is limiting equipment in terms of hydraulic capacity, the filtration capacity at the design temperature must be higher than the maximal instantaneous flow rate of water to be treated. Special attention should be paid to the peak factor and also to the parasite water flow rate that will be admitted at the treatment station.

Table 1.6 presents the usual values observed in MBR stations by various suppliers of membranes.

	Hollow fibers	**Flat sheet membranes**
Instantaneous flux at 20°C ($L.m^{-2}.h^{-1}$)	20	30
TMP (bar)	0.1–0.4	0.04–0.15
Permeability ($L.m^{-2}.h^{-1}.bar^{-1}$)	80–150	200–800
Specific membrane aeration demand ($Nm^3.h^{-1}.m^{-2}$)	0.2–0.5	0.2–1

Table 1.6. *Operating parameters*

Membrane technology was implemented in the treatment of wastewater due to its high separation properties. Acting as a physical barrier between two phases, a motive force is needed to obtain a flux. The transmembrane flux can be caused by a temperature gradient, a concentration gradient or a hydraulic pressure gradient. The motive force in the membrane processes of water treatment is generally the hydraulic pressure gradient, also known as TMP. It is equal to the difference between the supply pressure and the permeate pressure. It is therefore the head loss through membranes.

TMP is an indicator of membrane clogging at a given moment and it is highly dependent on the permeate flux, on temperature, on the filterability of biomass and on concentration. A too high TMP accelerates the clogging and can damage membranes. Its value is determined by calculating the absolute value of the difference between the pressure read upon the shutdown of the installation (the circuit being initiated) and the dynamic pressure. As an example, TMP in the filtration phase ranges between 100 and 300 mbar for the hollow fiber membranes and between 50 and 100 mbar for the flat sheet membranes.

By convention, it is negative in the filtration phase (permeate suction in the membranes) and positive in the backwashing phase (permeate backflow in the membranes):

$$\text{TMP} = \text{Abs (P inlet} - \text{P outlet)}$$

In the MBR with external loop, the TMP is obtained by the equation (Cardot 1999):

$$\text{TMP} = \frac{\text{Pa} + \text{Pc}}{2} - \text{Pp}$$

In this equation, Pa is the supply pressure, Pp is the permeate pressure and Pc is the concentrate pressure. Filtration operations can be conducted either by imposing a pressure gradient or a permeation flux. MBRs must always be operated below a certain TMP known as the critical pressure. The value of the critical pressure depends on the chemical nature of the membrane and is often provided by the supplier. Most installations operate at a constant flow rate, and a suction pump is used for immersed membranes, even though TMPs of several tenths of a bar are already ensured by the levels of water present in the tanks or by the level of discharge of the permeate. The flux is a crucial parameter in the evolution of TMP. Even though filtration is conducted below critical conditions, a change in TMP is observed.

Permeability (Lp20) is an indicator of long-term stability of filtration performances. It corresponds to the filtration capacity per m^2 and per bar. A high permeability indicates a good general state of membranes and corresponds to a lower TMP. It is always calculated at 20°C, on the basis of the instantaneous flux:

$$\text{Lp20} = \frac{\text{Flux}}{\text{TMPfiltration}} e^{-0.0239(T-20)}$$

With the temperature correction:

$$\text{Lp20} = \frac{\text{Ji(20)}}{\text{TMPfiltration}} = \frac{\text{Ji(20)}}{\Delta \text{Pf}}$$

Permeability is expressed in $L.m^{-2}.h^{-1}.bar^{-1}$ and more rarely in $m.d^{-1}.bar^{-1}$. As an example, a new membrane with hollow fibers has a sludge permeability $\geq$ 500 lmh/bar at 20°C.

Nevertheless, clean water permeability is not as significant in MBR processes as in conventional processes, as the accumulation of clogging particles at the level of the surface will strongly impact the properties of membrane filtration.

Permeability is mainly governed by the cross-flow velocity of the flux in the MBR with an external loop and by aeration in the immersed MBR. In the latter, air injection under and inside the membrane attenuates the formation of a cake on the external surface of the membrane (Pellegrin et al. 2002; Matosic et al. 2008). Aeration produces the oxygen needed for the biological activity and creates turbulence around the membrane which removes the large particles/grosses and dislodges the cake formed under the effect of suction required for filtration. However, aeration may have negative impacts on the bioreactor performances. Alternating aeration may limit clogging.

While membrane performances are a critical factor in this process, the configuration of membrane modules, the design of processes, the operating conditions and the biological performances are even more essential.

However, membrane performances are sensitive to certain factors:

– dependence on flux and pressure when a cake is formed around the membrane;

– lack of differentiation between the resistance of the membrane and the resistance to fouling after a certain operating time;

– calculations based on the total surface of the membrane and not on the effective surface of the membrane.

Consequently, the measures of the fouling rate and of the drop in permeability can be considered a consequence of the degradation of the filtration process but their causes are difficult to explain.

The maximal temperature admitted around the membranes is 35°C. Beyond this value, membranes and potting (material that maintains fibers) may suffer serious damage.

The membrane tank is designed to receive chemical reagents for membrane cleaning. It must be able to bear/supporter the acidic, basic and oxidizing character. A 1% slope of the bottom of the tank allows its full drainage during cleaning operations. A vessel/caisson or a cesspit will be installed at the lowest point of the tank to allow complete drainage of the membrane tank. The membrane tank is designed to allow jet cleaning of aerators and of the bottom of the tank when lifting modules would not be possible.

The distribution of the mixed liquor in the membrane tank is an important parameter to be taken into consideration to avoid the differences of concentration in the tank, which would lead to a clogging disparity in the modules. Membrane aeration creates a strong turbulence around the membranes and contributes to a good homogenization of the sludge. However, dead zones in the membrane tank can lead to excessive accumulation of sludge on the elements and on the bottom, which is difficult to remove during exceptional cleanings.

In the presence of large membrane tanks, several inlet points of the mixed liquor are necessary to allow good homogenization of the MLSS concentration in the membrane tank. A pipeline is then installed on the bottom of the membrane tank to distribute the mixed liquor throughout the length of the tank. This pipeline having holes directed downwards also allows avoiding an accumulation of sludge at the bottom of the membrane tank.

Membrane aeration has a double objective:

– avoid any sludge overconcentration at the level of membranes, by creating an airlift;

– limit the physical clogging of membranes by preventing sludge adherence to membranes.

It is implemented using a forced draught of air at the base of the membrane modules through aerators more or less integrated with each element, depending on suppliers, in order to create turbulence that puts into suspension the SS accumulated at the surface of membranes. In such cases, an aeration rate of 0.2–0.9 $Nm^3.h^{-1}$ per square meter of membrane surface is generally recommended (Krause et al. 2010). Membrane relaxation periods can also be scheduled. Despite these measures, membranes must be periodically cleaned by backwashing with filtered water or chemical products, or by immersion in a bath of chemical products.

Membrane aeration is essential for obtaining good filtration performances. Membrane air production is ensured by one or several boosters, backflowing into one or several collectors shared by the set of modules of a row/file.

Membrane aeration can be applied between several modules or several groups of modules and can be either cyclic or continuous, depending on the supplier. In case of cyclic aeration, the choice of automatic valves should be properly studied. The number of times these valves are used should also be clearly stated in the specification.

During night time, MBRs can be shut down because of lack of water. Membrane aeration can be reduced to save energy for this station.

A permeate storage tank is necessary for the industrial water needs of the plant (screens, sludge treatment equipment). This tank can also be used for storing the permeate used during backwashing, degasification and maintenance cleaning.

Figure 1.26. *Non-aerated membrane tank*

The MLSS concentration around membranes must be limited at 12–14 $g.L^{-1}$ for hollow fiber membranes and at 18 $g.L^{-1}$ for flat sheet membranes, in order to limit the risks of membrane solidification. In case of configuration immersed in a separate tank, the MLSS concentration in the biological tank must be approximately 10 $g.L^{-1}$ for hollow fibers and 14 $g.L^{-1}$ for flat sheet membranes.

In the case of seasonal loading variations, sludge age must be limited to 80 days in the low season. One method to limit sludge age involves using a tank whose volume depends on the season. To control sludge age, the station can be operated at various MES concentrations depending on actual needs.

1.6.3. *Management of membrane filtration*

MBR stations rely on a management system based on the control of operating conditions throughout the system. Flux demands are defined by the ingoing flux, and each downstream treatment phase must comply with the demands defined by the previous phase. In classical stations, control is generally facilitated by the process of overflow to the next phase. On the other hand, the adaptation of performances of the membrane system requires time. This is why a buffer work should be considered.

For urban applications, a 15 minutes buffer is recommended to allow the filtration unit to pass from minimum flux to maximum flux.

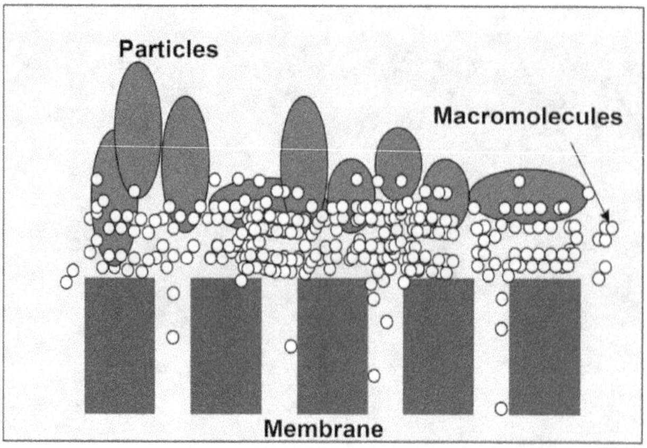

a)

b)

Figure 1.27. *Membrane clogging: a) simplified diagram; b) in station*

Beyond the membrane filtration, an MBR station is monitored similarly to a classical activated sludge station. But it is essential to implement a tool aiding efficient decision-making to simplify the operation in terms of filtration performances, particularly in the presence of several filtration rows/files of filtration. Data analysis requires a high-performance system to ensure a proper monitoring of the membrane performances.

To avoid excessive clogging of membranes, the pressure differential between upstream and downstream of the membrane must be continuously monitored and connected to alarms. Membranes can also be subjected to a loss of integrity. If needed, a device may be installed to allow leakage detection. All mechanical equipment and accessories required for membrane cleaning, MBR monitoring and maintenance must be included in the project.

Veolia developed its own software for data acquisition and processing. As part of the Veolink Care package, it is very simple to implement on site and is used in most Biosep® projects (Veolia MBR).

The control of MLSS leakage in the permeate circuit is essential for the efficiency of the treatment. The procedure consists of installing on the permeate circuit either a prefilter (control and prevention function) or an online turbidimeter (only control function). The objective of the prefilter is to prevent membranes from being clogged by SS that could have been sucked in filtration, such as, for example, in the case of breakage at the level of permeate connections.

It is recommended to install a prefilter with a cutoff of 80 µm, associated with a system of detection of high differential pressure connected to the optional automatic control system along the permeate circuit. In case of detection, filtration cycles are stopped. The prefilter is installed downstream of the membranes in order to detect any SS leakage and prevent the SS pumped return to the membranes during backwashing.

The online turbidimeter should be of the latest generation with a good resolution to be able to detect very fine particles. It is essential to install the turbidimeter with a bubble trap upstream to avoid signal perturbations due to microbubbles.

1.7. Membrane clogging

Nowadays, the main factor limiting the long-term application of MBRs is the clogging of membranes. Seyhi (2011) reports that by splitting clogging according to three observation scales – macroclogging (SS accumulation on the membrane), microclogging (development of biofilm on the membrane) and nanoclogging (adsorption of molecules) – the classification includes:

– macroclogging, underlining the importance of activated sludge and its role in clogging (Li et al. 2005), with SS considered as the main cause of membrane clogging (Yamamoto et al. 1989; Brindle and Stephenson 1996). The interest of this cake, as a selective dynamic layer, is that it can improve the retention capacity of the membrane (Tardieu et al. 1999; Ognier et al. 2004), but will rapidly become limiting depending on the permeate flux (> 10 lmh);

– microclogging, highlighting the importance of soluble microbial products (SMPs) generated during biological reactions and extracellular polymeric substances (EPSs) (Defrance 2000; Lee et al. 2003);

– nanoclogging, with a significant contribution of dissolved compounds. According to Fane et al. (1980), dissolved substances contribute up to 60% to clogging. This membrane clogging must be managed and controlled in order to maintain the permeability conditions required for membrane longevity and system productivity. Customized operating conditions and design can actually minimize macroclogging (Lebegue et al. 2009). Among other results, clogging management leads to reducing the costs related to membrane cleaning operations.

1.7.1. *Parameters influencing membrane clogging*

Several studies tried to clarify the mechanisms and factors that influence and govern membrane clogging. Even though clogging is the result of complex phenomena, it is significantly influenced by the operating and hydrodynamic conditions, the membrane type and configuration and the sludge characteristics (SS, physiological state of biomass, viscosity, etc.) (Cicek et al. 1998; Choi et al. 2005; Kaam et al. 2006). It is therefore a complex phenomenon. It can be described as being a continuous decrease of the permeate flux (under constant pressure) or an increase of TMP, caused by the accumulation or adhesion and/or adsorption of soluble or colloidal (organic, inorganic and microbial cells) particles or particles in suspension at the surface of the membrane and on the pores of the membrane (Choo and Lee 1996). It is possible to quantify clogging depending on its reversibility: reversible clogging (of a physical nature) and the irreversible clogging (of a chemical nature). Several factors can contribute to membrane clogging, and the description of this clogging varies from one author to the other, but it is rather the reflection of operating conditions and of the membrane configuration used.

The wide majority of membranes used in MBRs are based on polymer. A direct comparison between polyethylene (PE) and PVdF membranes clearly shows that PVdF membranes are less sensitive to irreversible clogging.

The type of material of organic membranes must be:

– sufficiently mechanically and chemically robust to bear the constraints imposed during filtration and cleaning cycles;

– easily modifiable to provide a hydrophilic surface, which then renders it more resistant to fouling, especially by exopolysaccharides (EPSs), and it is easily linked to a substrate to provide sufficient mechanical integrity.

Concentration polarization as well as clogging are considered major limits that tend to reduce membrane performances. The concentration polarization phenomenon is the consequence of the accumulation of compounds in the vicinity of the membrane. A decrease in the TMP or a filtration shutdown allows the instantaneous removal of the concentration polarization layer: the phenomenon is therefore reversible. As for the clogging, there are three ways in which it can occur. It may be from the adsorption of compounds on the membrane material with which there are strong interactions. Deep clogging may also occur when the size of compounds on the filter is smaller than that of the membrane pores (case of macromolecules present in the MBR). The third form of clogging is the deposit of particles at the surface of the membrane, which may generate a more or less compact cake (Le-Clech et al. 2006). The clogging of MBRs is generally a combination of these three forms. It depends on the physical characteristics of membranes, on those of used water and on the operating conditions.

The pore diameter makes it possible to distinguish between a microfiltration membrane and an ultrafiltration membrane and addresses the importance of discharge and the control potential of the organic and bacterial clogging. Indeed, the effects of the diameter of pores and of their distribution are strongly related to the characteristics of the effluent to be treated and of the bacterial populations present in the bioreactor. If the size of particles is smaller than the diameter of pores, obstruction of pores and limitation of the treatment flux will be rapidly observed. For this reason, some providers of membranes prefer to use a microfiltration membrane rather than an ultrafiltration membrane.

This parameter remains, however, insufficient to predict the hydraulic performances given the complexity of the biological and particulate matrix in the reactor. Other factors are involved, such as the hydraulic resistance of the membrane, the transmembrane velocity, the applied pressure and flux. Fortunately, a part of the clogging is reversible and is easily removed during the maintenance cleaning.

Membrane roughness and porosity are involved in the membrane clogging mechanisms. A membrane with dense structure and low porosity is more sensitive to clogging. For example, polyvinylidene fluoride (PVdF) and polyethersulfone (PES) membranes with pores of similar diameters may have different fouling behaviors. While fouling is mainly due to the cake surrounding PVdF and AC membranes, the blocking of pores is responsible for over 80% of the total hydraulic resistance of the PES membrane. Moreover, less fouling is observed with membranes having elliptical pores rather than circular pores.

The effect of roughness with respect to fouling of membranes shows that the higher the roughness, the more significant is the potential to accumulate fouling layers.

Due to hydrophobic interactions between the effluent, the bacterial biomass and the material of the membrane, membrane fouling is more serious with hydrophobic than with hydrophilic membranes. For example, the measurement of the contact angle showed that the apparent hydrophobicity of polyethersulfone (PES) decreased (from 55° to 47°) with the increase in the molecular weight cut-off (MWCO) parameter of organic materials passing from 20 to 70 kDa.

The loss of permeability can be explained by the dynamic layer formed in the cake deposited on the membrane, and, to a lesser extent, by the direct adsorption in the pores of the membrane and on its surface.

In MBRs, depending on the sludge age, activated sludge may contain significant quantities of hydrophilic EPSs (exopolysaccharides), which are known for their significant contribution to membrane clogging.

The TMP that measures the difference in pressure between the two compartments of the membrane is an important parameter for the evaluation of performances of MBR systems as it allows preventing membrane clogging. Indeed, this clogging phenomenon leads to an increase in hydraulic resistance and, at the same time, to an increase in TMP and to a decrease in the permeate flux when the system operates, respectively, under constant PTM or flux. The drop in flux or the increase in TMP requires the membrane to be hydraulically or chemically regenerated at regular intervals. For example, with a sludge concentration (MLSS) in the biological tank of 10 $g.L^{-1}$, TMP increases progressively to reach, at the end of 20–40 days, the limit value of clogging fixed at 500 mbar. At this moment, the activity of biomass is intense and results in the formation of a sludge cake that adheres to the membrane pores, blocks the filtration operation and drives an increase in TMP. Clogging is the vulnerability of membrane technology and can be defined as an undesirable deposit and an accumulation of microorganisms, colloids, cell debris and other solutes either inside the pores or at the surface of membranes.

The presence of an airlift may delay clogging. TMP becomes maximal after 50–60 days of operation.

In MBRs, it is important to anticipate the prediction of the resistance to fouling for the optimization of the operating conditions. Among these strategies, the membrane geometry and the modification of the surface to reduce the biological fouling are the object of special attention. In particular, the characteristics of the membrane, such as the material of the membrane, the size of pores and the hydrophilic character, are important factors for membrane fouling, as they affect the interactions between the surface of the membrane and the mixed liquor in the bioreactor. Hydrophilic membranes allow a higher attenuation of the membrane fouling than hydrophobic membranes.

Once clogging starts, membranes must undergo a chemical cleaning that is essential to prevent the irreversible clogging of membranes.

1.8. Membrane cleaning

The major problem of the membrane process is related to the continuous clogging of membranes in the filtration phase. While the purpose of membrane aeration is to avoid this clogging process, regular cleaning must be practiced in order to recover a part of the membrane performances. Adsorption-based clogging requires chemical cleaning to regenerate the filtration performances of the membrane. Pore blockage and the deposit of particles can be removed by the backwashing when membrane technology allows it (the case of hollow fiber membranes). For the other membrane technologies, chemical cleaning is required. Chemical washing involves using an acidic solution (e.g. citric acid) to remove the clogging due to inorganic material and an oxidizer (e.g. a NaOCl solution) to oxidize the organic material.

Permeability corresponds to the treatment capacity of an MBR brought to identical conditions of TMP and temperature. TMP is related to the quality of the membrane and to the quality of sludge. This is why, at equal flux, permeability evolves depending on the sludge age and on the clogging of membranes along the filtration cycles (see Figure 1.28).

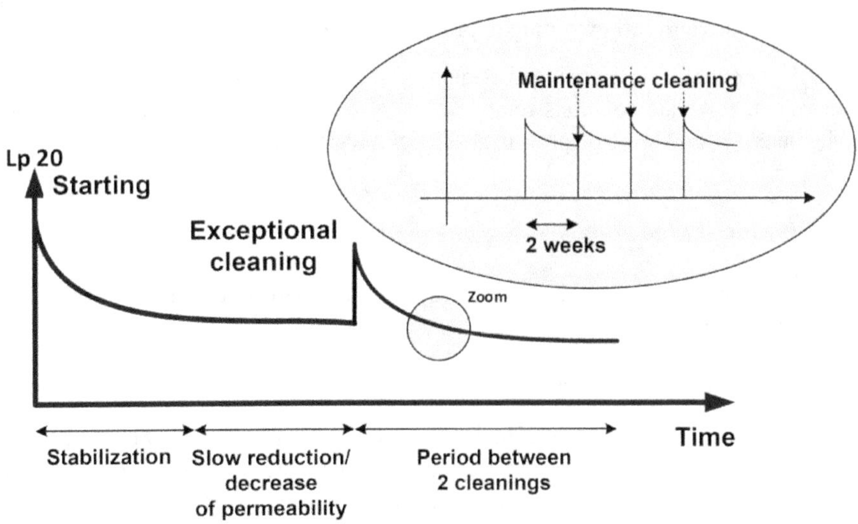

Figure 1.28. *Typical evolution of membrane permeability in time*

Membrane cleaning removes clogging and thus extends the life time. There are several techniques for cleaning membranes.

There are two ways to deal with fouling, which is the main obstacle to filtration:

– By mechanical and hydraulic action:

- Immersed membranes: fibers are permanently stirred by large air bubbles directly injected under the modules. Due to the resulting turbulences, the deposit formation is perturbed. It is also rinsed by backwashing with water, in the opposite direction to filtration.

- Tangential filtration: the effluent circulates at high velocity in the circulation loop and allows the evacuation of accumulated sludge.

– By backwashing and chemical washing, which are the most commonly used ways to remove fouling:

- Backwashing is conducted with water or with air. The larger the pores of the membrane, the less efficient is the backwashing. Several parameters influence the efficiency of backwashing, particularly the backwashing flow rate (in general, this flow rate is twice to three times higher than the filtration flow rate), the frequency and the duration of backwashing. The literature reveals that the shorter the backwashing, the more it contributes to reestablishing a better flux (Ramirez 1998; Barrios 2006). As only hollow fiber membranes can be backwashed (pull-out of adhesive lines and swelling of flat sheet membranes during backwashing), manufacturers are currently working to produce backwash flat sheet membranes.

- By chemical action: chemical cleaning can be used to attack the deposits fouling the membranes. Acidic solutions remove mineral deposits. Alkaline solutions remove organic materials, and chlorine prevents bacterial development.

1.8.1. *Chemical cleaning*

Two categories of chemical cleaning can be identified: maintenance cleaning and exceptional cleaning.

1.8.1.1. *Maintenance cleaning*

The maintenance cleaning procedure depends on the membrane provider. For the immersed membranes, maintenance cleaning involves backwashing with a chemical solution, generally with chlorine water or from time to time with citric acid (case of high hardness effluents). Maintenance cleaning is conducted in the opposite direction to filtration.

This cleaning is based on the injection of chemical products such as chlorine and/or acid in the membranes on the permeate side. The counter-pressure and the flow rate of the chemical injection vary depending on the supplier and on the membrane type. The frequency of the maintenance cleaning is a function of the type of influent. The current values are the following: 0.5–2 times/week for domestic wastewater.

The concentration of the chemical product depends on the membrane supplier. The pH specifications are also provided by the suppliers to enhance the efficiency of the chlorine cleaning. The pH limits specified by the membrane suppliers must be strictly observed.

A space should be dedicated to the on-site preparation of chemical products in order to host tanks for the storage of chemical products and the associated pumps. Chemical products must generally be diluted with a bit of permeate to reach the appropriate concentrations. The operation is conducted directly in the permeate storage tank. The chemical products are injected, by pumping or gravity, in the membranes on the permeate side. Flat sheet membranes do not bear the counter-pressure and therefore they generally need an intermediary storage tank that is slightly above the membrane modules, allowing the slow percolation of the chemical solution through the membrane.

After the chlorine cleaning, dechlorination and neutralization of residual waters are necessary in all cases, particularly if they are recycled at the head of the station ($6 < pH < 9$; free residual Cl < 0.05 $mg.L^{-1}$).

In all cases, the function of maintenance cleaning is to maintain the low permeability slope within an acceptable limit. Conversely, exceptional cleaning allows the recovery of a certain permeability when a low threshold of permeability was reached.

1.8.1.2. *Exceptional cleaning*

Some membrane suppliers recommend a more complete and deep cleaning. This cycle, referred to as exceptional cleaning or regenerative cleaning, involves submerging the membranes in a cleaning solution (chlorinated solution, acid or detergent). In general, this starts by a manual cleaning of membranes in order to remove the accumulation of sludge and fibers. Immersion periods can last up to 12 h, which explains the long duration of regenerative cleaning. The frequency of exceptional cleaning depends on the type of influent. Its common value is one or two times/year in urban wastewater.

Acidic solutions are mainly used to remove inorganic deposits, basic solutions and surfactants to remove organic materials and disinfectants to prevent the growth of microorganisms in the membrane system (Bouchard et al. 2000).

Chemical products are chosen depending on the nature of the membrane material, in order to respect the limits of pH and temperature as well as the tolerance to chlorine, other disinfectants and acids. Under normal conditions, the generally used solutions for chemical cleaning are sodium hypochlorite (NaOCl) to dislodge organic materials and citric acid ($C_6H_8O_7$) for inorganic materials (Lee et al. 2003; Chang and Kim 2005; Le-Clech et al. 2006).

Membrane manufacturers frequently propose scientific protocols specific to chemical cleaning. Most of these cleanings are operational without emptying the bioreactor where the cleaning solutions are sent during backwashing and have little impact on the biological process.

The concentration of the chemical product used depends on the membrane supplier and is generally higher than that related to maintenance cleaning. The procedures are, however, different for immersed membranes depending on whether it is a single tank or a separate tank.

1.8.2. *Implementation of cleaning*

1.8.2.1. *Single-tank configuration*

The membrane module is taken out of the tank by means of a cleaning device; the operator must wash it manually in order to clean it and remove the accumulated sludge. A low pressure water jet is used to prevent membrane damage. After the manual cleaning, the module is dived in a cleaning tank filled with a chemical solution at a given concentration for a given period of time (variable depending on the temperature of the cleaning solution).

1.8.2.2. *Separate-tank configuration*

For the separate-tank configuration, the exceptional cleaning protocol is, in principle, similar to the one for the single-tank design. It involves extended soaking of all membranes of the same tank in at least two cleaning solutions (chlorine, acid), for an in-depth removal of the elements that may clog the pores. The modules of the membrane tank are not displaced, the membrane tank playing the role of cleaning tank. The mixed liquor is evacuated from the membrane tank to the biological tank by pumping. A tidal range should be considered in the biological tank, as well as a pipeline system to allow sludge evacuation from the membrane tank to the biological tank.

The exceptional cleaning in a separate tank involves the following six phases:

1) *Drainage of the membrane tank*: the mixed liquor is evacuated by pumping to the biological tank.

2) *Rinsing*: before introducing the cleaning solution, membranes are rinsed. For this, membrane aeration is maintained during the sludge drainage up to an intermediary threshold (50 cm above aerators). The membrane tank is filled with permeate coming from another row/file in the case of a configuration with several rows/files (a meshing on the permeate collectors should be considered) or a permeate tank. Backwashing is conducted in order to again suspend the sludge stuck to the fibers (with the aeration). The rinsing solution is drained, by gravity or by pumping to the "all-water" station.

3) *Chlorine cleaning*: a chlorine solution at a given concentration is introduced in the membrane tank for a given duration (typically 12–24 h). The cleaning solution is then drained, to the all-water station or to the biological tank if the tidal range volume allows it, and if the chlorine quantities do not endanger the biology (total Cl_2 dose below 4 $gCl_2.kg^{-1}$ VSS $.d^{-1}$) (VSS: volatile suspended solids).

4) *Acid cleaning*: an acidic solution at a given pH is introduced in the membrane tank for a given duration (typically 3–6 h).

5) *Acid neutralization and drainage*: the acid solution is neutralized by the addition of caustic soda in the membrane tank before draining it to the "all-water" station. On the other hand, no neutralization is required if the solution is sufficiently dissolved in the biological tank (in other terms, the pH of the mixed liquor where the acid is injected should not be influenced by the acidic solution; the latter should therefore be injected where alkalinity is maximal in the treatment row).

6) *Restart*: when the drainage of the membrane tank is completed, the mixed liquor of the biological tank is sent to this tank up to the operating level corresponding to the restart of filtration.

Membranes with an external loop are cleaned by circulating through them an alkaline solution of sodium hypochlorite and an acidic solution, in the direction of filtration.

For urban wastewater, one or two exceptional cleanings are recommended per year. This figure can be increased depending on the operating conditions and on the quality of the wastewater to be treated.

1.9. Performances and warranties

The performances of MBRs are interesting both in relation to biodegradable and difficult to biodegrade pollutants and to particulate materials. MBRs also ensure an advanced disinfection whose intensity depends on the cutoff and on the distribution of the diameters of membrane pores. The treated effluent has excellent quality in terms of reduction of turbidity, SS, BOD, COD, disinfection and by-products of disinfection. This treated water is completely acceptable for direct reuse at the municipal level (toilet flush water, car washing, etc.) or for indirect use at the industrial level (cooling water, process water, etc.). It might also serve as feedwater for reverse osmosis (RO) units (Xing et al. 2000). The additional investment and operation cost of an MBR is then compensated by the minimization of its covered surface (by a factor of four with respect to a conventional activated sludge process) and the fact that the quality of water produced does not require a tertiary treatment such as oxidizing or infiltration/percolation.

1.9.1. *Physico-chemical and microbiological guarantees*

The synergistic filtration–biodegradation action allows the complete removal of SS. At the same time, the intense microbial activity at the level of the reactor allows a removal of organic material in terms of COD and BOD_5. The COD and BOD_5 residuals depend on soluble or biodegradable fractions of raw water. In general, the reduction rates of organic pollution in wastewater are very interesting. MBR is very robust to unexpected loading variations and may operate under significant volume and organic load. A 99% reduction of COD was observed on a wastewater of organic load of the order of 10 500 mg $COD.L^{-1}$ (Zhang et al. 2002). Several studies report COD reduction rates of up to 99%, depending on the type of treated water, organic compounds, membrane type and imposed treatment conditions (TMP, θ, etc.). The aeration, carbon availability and the pH of the medium facilitate nitrogen removal. However, the direct removal of phosphorus remains moderate and other techniques involving chemical reagents (iron salts) are used when the required residual phosphorus concentration is very low.

The guarantee for COD is based on the hard COD of raw water. A value of 30 $mg.L^{-1}$ is valid for standard French raw water. The efficiency of an MBR makes it possible to reach an annual or periodical average concentration below 5 mg $BOD_5.L^{-1}$ and below the detection limit of the SS method (3 $mg.L^{-1}$) or to ensure at least 97% and 99% removal, respectively, for these parameters.

MBRs are also used to remove nitrogen (N-NH_4 and N-NTK) and phosphorus (P-PO_4 and total P). It is then possible to use the recirculation flow rate of the membrane tank (300–400%) to return to the top the nitrates produced during the aerobic phase. Pt and N-NH_4 removal rates going up to 97% and 99% can be, respectively, obtained, depending on operating conditions and on the type of treated effluent. However, the application of MBRs with urban wastewaters requires a higher

θ (> 20 days), taking into account the complexity of the effluent. In a conventional MBR, denitrification is due to the presence of anoxia zones caused by high concentrations of biomass. Indeed, at a low concentration of dissolved oxygen, the limits of diffusion can create, in the aerobic tank, anoxia zones between biological flocs, thus facilitating denitrification (Pochanna et al. 1999). However, the simultaneous nitrification and denitrification processes require taking into account certain operating and MBR configuration conditions. The type of aeration may have a significant effect on the growth and development of nitrifying and denitrifying bacteria. The concentration of dissolved oxygen required for carbon oxidation and simultaneous nitrification must be greater than or equal to 2 mg $O_2.L^{-1}$ (Sorensen and Jorgensen 1993). Intermittent aeration can be used to reach this concentration. Nitrogen removal can also be improved with the creation of an anoxia zone separated from the aerobic zone (Côté et al. 1997). N-NH_4 and NT reduction rates above 90% and 97% are obtained. It should be noted that during an underload period, it is very difficult to control denitrification because of the oxygen supplied by the membrane air, sometimes exceeding the needs of O_2.

In general, MBR performances in terms of biological removal of phosphorus are low. A 45–67% reduction rate of phosphorus, by biological means, is observed (Vera et al. 1997; Battistoni et al. 2006). The degree of P removal depends on the (physico-chemical or biological) type of station.

There is practically no SS in the effluents exiting MBR. Consequently, the particulate fraction of N and P encountered at the outlet of conventional activated sludge stations is not found in the effluent.

Though each membrane supplier can propose its own process warranties, Table 1.7 indicates the warranties that are generally easy to reach provided the activated sludge process is properly designed.

However, several factors are involved in ensuring the best efficiencies. Certain factors are intrinsic and related to the biological medium and to the operating conditions: pH, concentration of dry material, aeration rate, permeation flow rate/relaxation time and sludge content. Other factors are inherent to the nature and composition of membranes.

The optimization of the operation of an MBR is based on the optimization of these factors, and this is an essential step for achieving an optimal membrane filtration. However, the limiting factor for membrane filtration remains the clogging of membranes which causes a loss of MBR performances by influencing TMP and permeate flux.

Parameters	Concentrations in the effluent
COD	≤ 30 mg.L^{-1}
BOD_5	≤ 5 mg.L^{-1}
SS	≤ 5 mg.L^{-1}
NT (total nitrogen)	Comparable to a classical activated sludge
PT (total phosphorus)	Comparable to a classical activated sludge

Table 1.7. *Process warranties for the MBR*

The disinfecting effectiveness of MBRs has been evaluated. MBR application allows the efficient removal of pathogen indicators (total and fecal coliforms), viruses and aerobic heterotrophic bacteria (AHB). This process also removes a very large part of pathogens and can be followed by a complementary disinfecting stage according to local regulations. The membranes used in MBRs have a nominal porosity ranging between 0.01 and 0.2 μm. This type of membrane can therefore retain bacteria such as coliforms, but does not seem able to retain all viruses with a diameter ranging between 0.001 and 0.1 μm. It is important to pay special attention to warranties related to the removal of microorganisms. Indeed, even though the membrane barrier may achieve a 5 log reduction of bacteria and a 1–4 log reduction of viruses, the residual may be higher than the concentrations required by the local regulation.

The disinfecting effectiveness of MBR is highly dependent on the size of pores. As the size of pores is never uniform, the selected membrane must have a sufficient gap with respect to the smallest microorganisms to be retained. However, microclogging reduces the cut-off threshold of membranes and some membranes, with a cut-off threshold of 0.4 μm, present a total bacteriological reduction.

1.9.2. *Removal of endocrine disruptors (EDs)*

Many endocrine disruptors (EDs) escape classical wastewater treatment. There are few studies on ED removal in MBRs. Most of them concern phenolic compounds (e.g. nonylphenol, bisphenol A), hormones and hormone derivatives (17β-ethinylestradiol, estradiol), several pharmaceutical compounds, etc.

The reduction rate of bisphenol A in an MBR generally oscillates between 93 and 97%. Nonylphenols, estradiol and 17β-ethinylestradiol are removed at a rate of 55%, 64% and $\geq 71\%$, respectively. Over 93% of genistin is removed.

Pharmaceutical products and derivatives (drugs, drug metabolites, cosmetic products, food supplements and derived metabolites) have also been highlighted, particularly the lipid regulator benzafibrate, analgesics (ibuprofen), polycyclic compounds (galaxolide and tonalide) and the antiepileptic carbamazepine, to cite but a few (Clara et al. 2005a, 2005b; Lyko et al. 2005; Chen 2008; Lee et al. 2008; Seyhi 2011).

According to study results, benzafibrate and ibuprofen have high removal rates ($\geq$ 95%). They are followed by polycyclic compounds (tonalide and galaxolide), which are removed at a rate of nearly 80%. Carbamazepine is not removed. Compounds such as phenazone and propyphenazone are partially (70%) removed in MBRs.

In tertiary treatment, aimed at reuse for industrial applications, MBR is often the treatment stage (Gaid 2025) before a final stage consisting of RO. The silt density index (SDI) must generally reach a value ≤ 3 upon entering RO, and this value can be normally reached after an MBR treatment.

1.9.3. *Hydraulic guarantees*

As the flux decreases with temperature, it is important to clearly specify the temperature ranges within which the hydraulic capacity can be guaranteed (the lowest temperature mentioned in the call for tender documentation). The filtration surface is dimensioned for the most penalizing case (peak duration and intensity, temperature). In a municipal context, the dimensioning case is most often the winter, when there are significant volumes and temperature is the lowest.

Moreover, the hypotheses considered for the dimensioning of the buffer tank (total peak duration) must be clarified in the guarantee specifications. A distinction should be made between a buffer tank and a storm water tank; the function of the first one is to store water reserves during maximum flux in the dry season and that of the second one is to store the over-flow of the water collected during rainy days.

As the flux decreases with membrane aging, design should take into consideration the customer guarantee duration.

1.9.4. *Membrane warranties*

Membrane lifetime is estimated between three and seven years depending on the system design and on the use of membranes, as the lifetime of membranes is directly related to their operating conditions. The warranty given by membrane suppliers may have various forms. It can be firmly offered for a reasonable duration (two years) by the supplier, and supplemented by a three-year *prorata temporis* warranty.

1.9.4.1. *Principles of* prorata temporis *warranty*

Under normal operating conditions as specified in the technical documentation (raw water quality, daily production, operating duration of the membrane system), and to the extent that the system is operated according to the manufacturer's recommendations, the customer will not need to replace the membranes for seven years, for example, if this is the period specified by the membrane supplier.

However, if a partial or total replacement of membranes is needed during the period of seven years, the costs are supported by the manufacturer and the customer according to the *prorata temporis* principle. This procedure is applicable if the manufacturing flow rate is not respected for a reason attributed to membranes (drop in permeability, abnormal breakage of fibers, etc.).

The *prorata temporis* warranty means that the warranty offered for the membranes is proportional to the operating time.

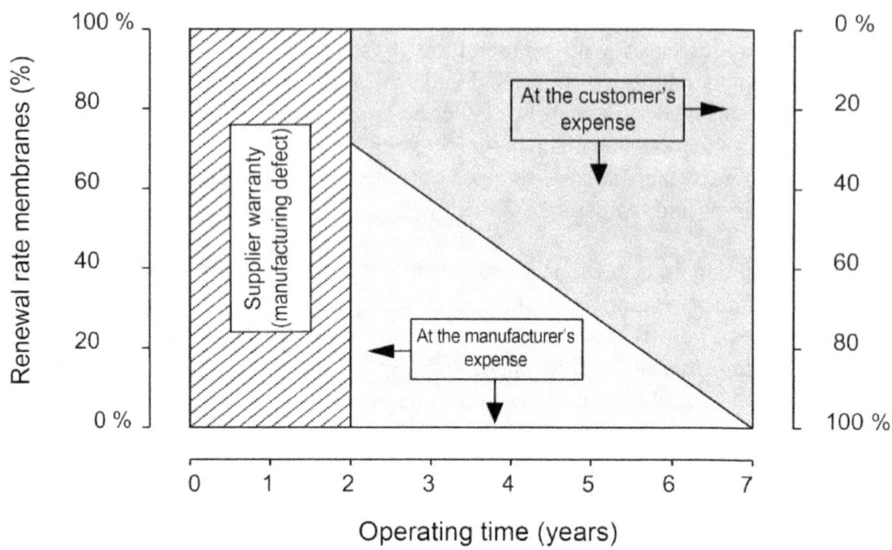

Figure 1.29. Prorata temporis *warranty operation*

The curve presented in Figure 1.29 shows the mechanism for prorating the membranes for the first operation cycle:

– The *prorata temporis* warranty comes into effect as soon as the plant starts operating, at the end of the period of observation.

– During the supplier warranty period which corresponds to the hatched area in Figure 1.29 (warranty against manufacturing defects), if the production flow rate is not reached because of defective membranes, the replacement of these membranes is at the supplier's expense. There are no charges for the customer.

– The *prorata temporis* calculation becomes applicable when the supplier warranty period expires. It involves customer participation in the replacement of membranes depending on their operating time (days, months, years). This procedure is implemented if the production flow rate of the plant is not met for a reason attributable to membranes (drop of permeability, abnormal breakage of fibers, etc.).

– After seven years of operation, the replacement of all membranes will be at the customer's expense. They will have benefited from drinking water production according to the technical specifications during this period.

– A new operating cycle is implemented according to the warranty procedure.

EXAMPLES.–

1) If at the beginning of the third year observations reveal that membranes are the permanent cause of a drop in production, the manufacturer (and the supplier) will be responsible for the replacement of 100 x (7 – 2)/7) = 71.4% of membranes, and the customer will be responsible for the remaining 28.6% because of satisfactory operation of the plant for two years and the customer's benefit of drinking water production according to the technical specifications during this period.

2) If these incidents occur at the end of the fifth year, the manufacturer (and supplier) will be responsible for 100 x (7 – 5)/7 = 28.6% of the replacement of membranes, and the customer will be responsible for the remaining 71.4% because of satisfactory operation of the plant for five years and the customer's benefit of drinking water production according to the technical specifications during this period.

1.10. Example of dimensioning of MBR with external loop

1.10.1. *Characteristics of supply water*

The treatment plant receives raw wastewater coming from domestic networks. The raw wastewater entering the plant is pretreated via screens and sand and grease removal before getting into the bioreactor. Wastewater characteristics are presented in Tables 1.8 and 1.9.

Given the high peak factor, the installation must buffer a part of the ingoing peak flow rate. The pretreatment is expected to take 1,770 $m^3.h^{-1}$ before the buffer tank (fine screening and grit removal-degreasing) and 1,070 $m^3.h^{-1}$ after the buffer tank (screening).

In order to comply with the specified maximal peak flow rate, a buffer tank of a net volume of 1,500 m^3 is integrated in the design.

The load considered for the design of the plant corresponds to a maximum daily flow rate of 20,500 $m^3.d^{-1}$ multiplied by the concentrations.

Hydraulic parameters	**Average**	**Maximum**	**Notes**
Daily average flow rate ($m^3.d^{-1}$)	17,000	20,500	–
Hourly average flow rate ($m^3.h^{-1}$)	708	854	
Peak flow rate factor	–	2.5	Peak for max. two hours/day two times/day
Feed flow rate ($m^3.h^{-1}$)	708	1,770	Maximum for max-two hours/day two times/day

Table 1.8. *Hydraulic parameters*

Wastewater parameters	**Base**	**Pollutant load (kg.d^{-1})**
Temperature (°C)	22–35	–
Temperature air (°C)	10–60	–
pH	6–9	–
COD (mg O_2.L^{-1})	500	10,250
BOD_5 (mg O_2.L^{-1})	250	5,125
Total SS (mg.L^{-1})	250	5,125
Total Kjeldahl nitrogen (mg N.L^{-1})	55	1,127
Total phosphorus in P (mg P.L^{-1})	8	164
Oil and grease (mg.L^{-1})	40	820

Table 1.9. *Raw wastewater parameters*

1.10.2. *Pretreatment required for the MBR*

A 6 mm raw screening followed by a 1 mm fine screening are required.

Parameters	Average values Raw water	Maximal values Raw water
Temperature (°C)	22–36	22–36
MES concentration in the biological tank (g.L^{-1})	11.3–36°C 12.7–22°C	15
Filtrability	See the rest of the section	
%VSS in sludge	60–70%	–
Sludge age	Minimum 17 days	–

Table 1.10. *MBR operating conditions*

Operating conditions	Operating flow rates		
	At 636 m³.h^{-1}	**At 811 m³.h^{-1}**	**At 1 070 m³.h^{-1}**
Average permeate/skid flow rate (including washing)	31.8 m^3.h^{-1}	40.6 m^3.h^{-1}	53.5 m^3.h^{-1}
Applied flux (L.m^2.h^{-1})	40.7	52	68.6
Max permeate/skid flow rate	36 m^3.h^{-1}	46.9 m^3.h^{-1}	60 m^3.h^{-1}
TMP during filtration	0.10–0.19 bar	0.12–0.24 bar	0.15–0.30 bar
Recirculation/skid flow rate	390 m^3.h^{-1}	520 m^3.h^{-1}	520 m^3.h^{-1}
Pressure drop on skids	0.24 bar	0.26 bar	0.26 bar
Air/skid flow rate	115 Nm^3.h^{-1}	229 Nm^3.h^{-1}	229 Nm^3.h^{-1}
Applied air pressure	0.6 bar	0.6 bar	0.6 bar
Washing/skid flow rate	234 m^3.h^{-1}	234 m^3.h^{-1}	234 m^3.h^{-1}
Time between two washing processes	10 min	7 min	7 min
Washing/skid time	7 s	7 s	7 s
TMP on membranes during washing	0.5–0.8 bar	0.5–0.8 bar	0.5–0.8 bar

Table 1.11. *Operating conditions of membrane filtration*

Sludge filterability must be sufficient for the MBR process. Besides the sludge age, filterability will be evaluated by means of a laboratory test (C5 paper for 5 min) and should normally reach a reference value > 15 ml. Taking into account the peak flow rate of 1,770 m^3.h^{-1} two times and two hours per day, the permeate production of the membrane workshop consisting of 20 operating skids will be the following during the day:

– average case (17,000 m^3.d^{-1}):

- 2 h at 1070 $m^3.h^{-1}$,
- 10 h at 636 $m^3.h^{-1}$,
- 2 h at 1070 $m^3.h^{-1}$,
- 10 h at 636 $m^3.h^{-1}$;

– maximum case (20,500 $m^3.h^{-1}$):

- 2 h at 1070 $m^3.h^{-1}$,
- 10 h at 811 $m^3.h^{-1}$,
- 2 h at 1070 $m^3.h^{-1}$,
- 10 h at 811 $m^3.h^{-1}$.

The workshop is composed of 20 skids with an installed membrane surface of 780 m^2 (26 modules with a unit surface of 30 m^2 or a total surface of an installed membrane of 15,600 m^2).

Due to the aeration of membrane modules, the level of oxygen dissolved in the tank increases generally from 1.5 to 2.5 $mg.L^{-1}$ during the passage in the membrane installation.

Chemical cleaning must be conducted every 30–120 days (depending on the quality of the bioreactor sludge).

A chemical cleaning is conducted with citric acid followed by sodium hypochlorite (NaOCl) for each individual skid. This skid is taken out of service during this cleaning. It is recommended to conduct chemical cleaning when the flow rate is the lowest (during night).

Chemical agent	Contact time	Interval	Concentration of reagents
Citric acid	30–120 min	30–120 d	10 $g.L^{-1}$
Sodium hypochlorite			400–500 ppm free chlorine

Table 1.12. *Conditions for the chemical cleaning of membranes*

1.10.3. *Characteristics required for the treated water*

The membrane workshop will meet the required warranties provided that the bioreactor operates according to the characteristics indicated in Table 1.13.

Parameters	Maximal concentration
COD	$\leq$ 50 mg.L^{-1}
BOD_5	$\leq$ 10 mg.L^{-1}
SS (mg.L^{-1})	$\leq$ 2
Turbidity (NTU)	$\leq$ 0.5
pH	6–9
N-NH_4	$\leq$ 1 mg.L^{-1}
Total nitrogen (NGL)	$\leq$ 30 mg.L^{-1}
Dissolved oxygen	$\geq$ 2 mg.L^{-1}
Fecal coliforms	$\leq$ 5 MNP/100 ml

Table 1.13. *Guaranteed quality of the effluent outgoing from the MBR*

Guarantees are given over 95 percentiles of a daily sample.

Figure 1.30. *Quality of the filtered water*

1.11. Conclusion

The MBR system is a revolutionary technology in the field of water treatment. It is characterized by interesting performances in terms of removal of biodegradable pollutants and pollutants considered difficult to biodegrade by the usual processes of wastewater treatment. Waters treated by MBR have excellent quality in terms of SS and disinfection. MBR technology is relatively more expensive than the standard active sludge systems, but enormous efforts have been undertaken (significant reduction of membrane costs and operation costs) to ensure its industrial and commercial development. The MBR operating range (sludge age and MLSS content) is rapidly limited as a consequence of stirring and mixing problems and (oxygen) mass transfer. Due to excessive energy consumption during MBR operation, MBR users have had to reexamine their objectives to finally adopt an SRT ranging between 20 and 30 days, and MLSS contents of 6–8 g.L^{-1}. However, so far the technology is mainly competitive for the treatment of industrial wastewater where effluents are highly loaded. For municipal wastewater, the high energy demand of 30–50% (compared to the conventional activated sludge system) may limit the use of this technology. This limitation is reduced when considering the compactness of an MBR installation (four times reduction of the covered surface) and the process performances (comparable to a tertiary treatment). Moreover, many research teams are working to develop and optimize the MBR technology, in order to increase its competitiveness and adapt it to municipal applications.

In recent years, the costs have been reduced thanks to the progress in the domain of membrane manufacturing. Progress also led to an increase in the life time of membranes, a reduction in the costs of their replacement, a reduction in energy consumption with respect to the obtained permeate flux and a decrease in the operating costs due to the possibilities to reuse treated waters.

Nowadays, water users agree on the fact that the use of membrane technologies has gained its place in the wastewater treatment field. Indeed, membrane modules have many advantages compared to a conventional activated sludge treatment process. These advantages can be divided into three major categories: installation, operation and process. They are therefore more compact and the ground surface can be five times smaller. Membranes can be implemented in small spaces. Moreover, MBRs have the advantage of being modular. Due to these characteristics, this process can be readily implemented in the existing installations that already have biological reactors such as, for example, during the rehabilitation of old treatment plants where settling tanks can be reconverted into buffer tanks. The treatment of wastewater by MBR has many advantages concerning the removal of biodegradable pollutants and also of micropollutants considered to be difficult to degrade in the conventional systems for wastewater treatment. Several studies reported very high rates of reduction of organic and inorganic pollution depending on the type of treated waters, on the type of organic compounds, on the type of membranes used and on the imposed treatment conditions.

1.12. References

Atasi, M. (2006). *Membrane Systems for Wastewater Treatment*. McGraw-Hill, New York.

Baek, S.H. and Pagilla, K.R. (2008). Simultaneous nitrification and denitrification of municipal wastewater in aerobic membrane bioreactors. *Water Environ. Res.*, 80(2), 109–117.

Battistoni, P., Fatone, F., Bolzonella, D., Pavan, P. (2006). Full scale application of coupled alternate cycles-membrane bioreactor (AC-BRM) process for wastewater reclamation and reuse. *Water Pract. Technol.*, 1(4), 28–39.

Berland, J.M. and Juery, C. (2002). Les procédés membranaires pour le traitement de l'eau. Technical Document, FNDAE, No. 14, Ministère de l'Agriculture, de l'Alimentation, de la Pêche et des Affaires rurales, Direction de l'Espace Rural et de la Forêt, Paris.

Birima, A.H., Megat Mohd Noor, M.J., Muyibi, A.S., Idris, A. (2005). Simultaneous organic and nitrogen removal using anoxic-aerobic membrane bioreactor. *Int. J. Eng. Technol.*, 2, 36–42.

Bouchard, C.P., Kouadio, P.D., Ellis, D., Rahmi, M., Lebrun, R.E. (2000). Les procédés à membranes et leurs applications en production d'eau potable. *Vecteur environ.*, 33, 28–38.

Bouhabila, E.H., Ben Aim, R., Buisson, H. (1998). Microfiltration of activated sludge using submerged membrane with air bulking (application to wastewater treatment). *Desalination*, 118, 315–322.

Brindle, K. and Stephenson, T. (1996). The application of membrane biological reactors for the treatment of wastewaters. *Biotechnol. Bioeng.*, 49(6), 601–610.

Cardot, C. (1999). *Les traitements de l'eau : procédés physico-chimiques et biologiques. Cours et problèmes résolus*. Ellipses, Paris.

Chang, I.S. and Kim, S.N. (2005). Wastewater treatment using membrane filtration effect of biosolids concentration on cake resistance. *Process Biochem.*, 40(3–4), 1307–1314.

Chang, I.S., Lee, C.H., Ahn, K.H. (1999). Membrane filtration characteristics in membrane coupled activated sludge system: The effect of floc structure on membrane fouling. *Sepac. Sci. Technol.*, 34(9), 15–30.

Chang, H.-S., Chooa, K.-H., Leeb, B., Choia, S.-J. (2009). Review: The methods of identification, analysis, and removal of endocrine disrupting compounds (EDCs) in water. *J. Hazard. Mater.*, 172, 1–12.

Chee-Sanford, J.C., Arminov, R.I., Krapac, I.J., Garrigues-Jeanjean, N., Mackie, R.I. (2001). Occurrence and diversity of tetracycline resistance genes in lagoon and groundwater underlying two swine production facilities. *Appl. Environ. Microbiol.*, 67(4), 1494–1502.

Chen, J., Huang, X., Lee, D. (2008). Bisphenol A removal by a membrane bioreactor. *Process Biochem.*, 43(4), 451–456.

Cho, B.D. and Fane, A.G. (2002). Fouling transients in nominally sub-critical flux operation of a membrane bioreactor. *J. Membr. Sci.*, 209(2), 391–403.

Choi, H., Dionysios, D.D., Oerther, D.B., Sorial, G.A. (2005). Influence of cross flow velocity on membrane performance during filtration of biological suspension. *J. Membr. Sci.*, 248(1–2), 189–199.

Choo, K.H. and Lee, C.H. (1996). Membrane fouling mechanisms in the membrane-coupled anaerobic bioreactor. *Water Res.*, 30(8), 1771–1780.

Cicek, N., Winnen, H., Suidan, M.T., Wrenn, B.E., Urbain, V., Manem, J. (1998). Effectiveness of the membrane bioreactor in the degradation of high molecular weight compounds. *Water Res.*, 32(5), 1553–1563.

Cirja, M., Ivashechkin, P., Shaffer, A., Haffer, A., Corvini, P.F.X. (2008). Factors affecting the removal of organic micropollutants from wastewater in conventional treatment plants (CTP) and membrane bioreactors (BRM). *Rev. Environ. Sci. Biotechnol.*, 7(1), 61–78.

Clara, M., Strenn, B., Ausserleitner, M., Kreuzinger, N. (2004). Comparison of the behaviour of selected micropollutants in a membrane bioreactor and a conventional wastewater treatment plant. *Water Sci. Technol.*, 50(5), 29–36.

Clara, M., Strenn, B., Kreuzinger, N., Gans, O., Kroiss, H. (2005a). Removal of selected pharmaceuticals, fragrances and endocrine disrupting compounds in membrane bioreactor and conventional wastewater treatment plant. *Water Res.*, 39(19), 4797–4807.

Clara, M., Strenn, B., Kreuzinger, N., Gans, O., Kroiss, H. (2005b). The solids retention time – A suitable design parameter to evaluate the capacity of wastewater treatment plant to remove micropollutants. *Water Res.*, 39(1), 97–106.

Comeau, Y. (2006). Traitement tertiaire (polissage) du lisier de porc par un bioréacteur à membranes (BRM) immergées. Report no. 703035, École Polytechnique de Montréal, Montreal.

Côté, P., Buisson, H., Pound, C., Arakaki, G. (1997). Immersed membrane activated sludge for the reuse of municipal wastewater. *Desalination*, 113(2–3), 189–196.

Defrance, L. and Jaffrin, M.Y. (1999a). Réversibilité of fouling formed in activated sludge filtration. *J. Membr. Sci.*, 157(1), 73–84.

Defrance, L. and Jaffrin, M.Y. (1999b). Comparison between filtrations at fixed transmembrane. Pressure and fixed permeate flux: Application to a membrane bioreactor used for wastewater treatment. *J. Membr. Sci.*, 152(2), 203–210.

Defrance, L., Jaffrin, M.Y., Gupta, B., Paulier, P., Geaugey, V. (2000). Contribution of various constituents of activated sludge to membrane bioreactor fouling. *Bioresour. Technol.*, 73(2), 105–112.

Delgado, S., Diaz, F., Villarroel, R., Vera, L, Diaz, R., Elmaleh, S. (2002). Nitrification in a hollow-fibre membrane bioreactor. *Desalination*, 146(1–3), 445–449.

Dhaouadi, H. and Marrot, B. (2008). Olive mill wastewater treatment in a membrane bioreactor: Process feasibility and performances. *Chem. Eng. J.*, 145(2), 225–231.

Enegess, D., Togna, A.P., Sutton, P.M. (2003). Membrane separation applications to biosystems for wastewater treatment. *Filt. Sep.*, 40(1), 14–17.

Esplugas, S., Bila, D.M., Krause, L.G.T., Dezotti, M. (2007). Ozonation and advanced oxidation technologies to remove endocrine disrupting chemicals (EDCs) and pharmaceuticals and personal care products (PPCPs) in waters effluents. *J. Hazard. Mater.*, 149(3), 631–642.

Fan, F., Zhou, H., Hussain, H. (2006). Identification of wastewater sludge characteristics to predict critical flux for membrane bioreactor processes. *Water Res.*, 40(2), 205–212.

Fane, A.G., Fell, C.J.D., Nor, M.T. (1980). Ultrafiltration/activated sludge system – Development of a predictive model. *Polym. Sci. Technol.*, 13, 631–658.

Field, R.W., Wu, D., Howell, J.A., Gupta, B.B. (1995). Critical flux concept for microfiltration fouling. *J. Membr. Sci.*, 100(3), 259–272.

Gagnaire, J., Moulin, P., Marrot, B. (2008). Bioréacteurs à membranes : un intérêt grandissant. *Revue l'Eau, l'Industrie, les Nuisances*, 314, 37–47.

Gaid, K. (2025). *Wastewater Reuse*, vols 1 and 2. ISTE Ltd, London, and Wiley, Hoboken, NJ.

Gander, M., Jefferson, B., Judd, S. (2000). Aerobic MBRs for domestic wastewater treatment: A review with cost considerations. *Sepac. Purif. Technol.*, 18(2), 119–130.

Garcia, M.T., Campos, E., Dalmaul, M., Ribosa, I., Sanchez-Leal, J. (2002). Structure–activity relationships for association of linear alkylbenzene sulfonates with activated sludge. *Chemosphere*, 49(3), 279–286.

Gogate, P.R. and Pandit, A.B. (2004). A review of imperative technologies for wastewater treatment: Oxidation technologies at ambient conditions. *Adv. Environ. Res.*, 8(3–4), 501–551.

Grasmick, A., Cabassud, C., Spérandio, M., Winiewsli, C. (2009). Les bioréacteurs à membrane appliqués au traitement des eaux usées. *Gestion durable des déchets et des polluants*, W4140 [Online]. Available at: https://www.techniques-ingenieur.fr/base-documentaire/procedes-chimie-bio-agro-th2/gestion-durable-des-dechets-et-des-polluants-42495210/bioreacteurs-a-membranes-et-traitement-des-eaux-usees-w4140/.

Holakoo, L., Nakhla, G., Bassi, A.S., Yanful, E.K. (2007). Long term performance of BRM for biological nitrogen removal from synthetic municipal wastewater. *Chemosphere*, 66(5), 849–857.

Hu, J.Y., Chen, X., Tao, G., Kekred, K. (2007). Fate of endocrine disrupting compounds in membrane bioreactor systems. *Environ. Sci. Technol.*, 41(11), 4097–4102.

Jawad, H.A.R. (2008). Performance of water recycling technology. PhD Thesis, University of Wollongong, New South Wales.

Jefferson, B., Brookes, A., Le-Clech, P., Judd, S. (2004). Methods for understanding organic fouling in MBRs. *Water Sci. Technol.*, 49(2), 237–244.

Jeon, S., Rajabzadeh, S., Okamura, R., Ishigami, T., Hasegawa, S., Kato, N., Matsuyama, H. (2016). The effect of membrane material and surface pore size on the fouling properties of sub-merged membranes. *Water*, 8(602), 2–11.

Judd, S. (2008). The status of membrane bioreactor technology. *Trends Biotechnol.*, 26(2), 109–116.

Kaam, R.V., Anne-Archard, D., Alliet, M., Lopez, S., Albasi, C. (2006). Aeration mode, shear stress and sludge rheology in a submerged membrane bioreactor: Some keys of energy saving. *Desalination*, 199(1–3), 482–484.

Krause, S., Zimmermann, B., Meyer-Blumenroth, U., Lamparter, W., Siembida, B., Cornel, P. (2010). Enhanced membrane bioreactor process without chemical cleaning. *Water Sci. Technol.*, 61(10).

Le-Clech, P., Chen, V., Fane, T.A.G. (2006). Review fouling in membrane bioreactors used in wastewater treatment. *J. Membr. Sci.*, 284(1–2), 17–53.

Lebegue, L., Heran, M., Grasmick, A. (2009). Membrane air flow rates and HF sludging phenomenon in MBR. *Desalination*, 236(1–3), 135–142.

Lee, Y., Cho, J., Seo, Y., Lee, J.W., Ahn, K.H. (2002). Modelling of submerged membrane bioreactor process for wastewater treatment. *Desalination*, 146(1–3), 451–457.

Lee, W., Kang, S., Shin, H. (2003). Sludge characteristics and their contribution to microfiltration in submerged membrane bioreactors. *J. Membr. Sci.*, 216(1–2), 217–227.

Lee, J., Lee, B.C., Ra, J.S., Cho. J., Kim, I.S., Chang, N.I., Kim, H.K., Kim, S.D. (2008). Comparison of the removal efficiency of endocrine disrupting compounds in pilot scale sewage treatment processes. *Chemosphere*, 71, 1582–1592.

Lesage, N., Spérandio, M., Cabassud, C. (2005). Performances of a hybrid adsorption/submerged membrane biological process for toxic waste removal. *Water Sci. Technol.*, 51(6–7), 173–180.

Lesjean, B. and Huisjes, E.H. (2008). Survey of the European BRM market: Trends and perspectives. *Desalination*, 231(1–3), 71–81.

Lesjean, B., Ferre, V., Vonghia, E., Môslang, H. (2008). Market and design considerations of the 37 larger MBR plants in Europe. In *EDS MDIW08 Conference*, 20–22 October, Toulouse.

Li, X.Y. and Wang, X.M. (2006). Modelling of membrane fouling in a submerged membrane biore-actor. *J. Membr. Sci.*, 278(1–2), 151–161.

Li, X., Gao, F., Hua, Z., Du, G., Chen, J. (2005). Treatment of synthetic wastewater by a novel MBR with granular sludge developed for controlling membrane fouling. *Sepac. Purif. Technol.*, 46, 19–25.

Liang, S.L., Tao, T., Kekre, K.A., Seah, H. (2006). A modeling study of fouling development in membrane bioreactors for wastewater treatment. *Water Environ. Res.*, 78(8), 853–863.

Liu, R., Huang, X., Sun, Y.F., Qian, Y. (2003). Hydrodynamic effect on sludge accumulation over membrane surfaces in a submerged membrane bioreactor. *Proc. Biochem.*, 39(2), 157–163.

Liu, Z.H., Kanjo, Y., Mizutani, S. (2009). Removal mechanisms for endocrine disrupting com-pounds (EDCs) in wastewater treatment-physical means, biodegradation, and chemical advanced oxidation: A review. *Sci. Total Environ.*, 40(2), 731–748.

Lobos, J., Heran, M., Grasmick, A. (2009). Optimization of the operations conditions in membrane bioreactors through the use of ASM3 model simulations. *Desalin. Water Treat.*, 9(1–3), 126–130.

Lyko, S., Wintgens, T., Melin, T. (2005). Estrogenic trace contaminants in wastewater – Possibilities of membrane bioreactor technology. *Desalination*, 178(1–3), 95–105.

Marcano, J.G.S. and Tsotsi, T. (2010). *Membrane Reactors*. Wiley-VCH, Weinheim.

Marrot, B., Barrios-Martinez, A., Moulin, P., Roche, N. (2004). Industrial wastewater treatment in a membrane bioreactor: A review. *Environ. Prog.*, 23(1), 59–68.

Matosic, M., Vukovic, M., Curlin, M., Mijatovic, I. (2008). Fouling of hollow fiber submerged membrane during a long-term filtration activated sludge. *Desalination*, 219(1–3), 57–65.

Melin, T., Jefferson, B., Bixio, D., Thoeye, C., De Wilde, W., de Kroning, J., Van der Graaf, J., Wintgens, T. (2006). Membrane bioreactor technology for wastewater treatment and reuse. *Desalination*, 187(1–3), 271–282.

Metcalf & Eddy Inc. (2003). *Wastewater Engineering*. McGraw-Hill, New York.

Ministère de l'environnement (2012). Traitement des eaux usées. Report, Ministère de l'environnement, Quebec.

Ngan, L. and Kim, A.S. (2007). A mini-review of modeling studies on membrane bioreactor (MBR) treatment for municipal wastewaters. *Desalination*, 212(1–3), 261–281.

Ognier, S., Wisniewski, C., Grasmick, A. (2004). Membrane bioreactor fouling in sub-critical filtration conditions: A local critical flux concept. *J. Membr. Sci.*, 229(1–2), 171–177.

Orantes, J., Heran, M., Wisniewski, C., Grasmick, A. (2006). The influence of operating conditions on permeability changes in a submerged membrane bioreactor. *Sepac. Purif. Technol.*, 52(1), 60–66.

Parsons, S. (2004). *Advanced Oxidation Processes for Water and Wastewater Treatment*. IWA Publishing, London.

Pellegrin, M.L., Wisniewski, C., Grasmick, A, Tazi-Pain, A., Buisson, H. (2002). Sequenced aeration in a membrane bioreactor: Specific nitrogen removal rates. *Can. J. Chem. Eng.*, 80(3), 386–392.

Pochanna, K., Keller, J. Lant, P.J. (1999). Model development for simultaneous nitrification and denitrification. *Water Sci. Technol.*, 39(1), 235–243.

Ramirez, J.A. and Davis, R.H. (1998). Application of crossflow microfiltration with rapid backpulsing to wastewater treatment. *J. Hazard. Mater.*, 63(2–3), 179–197.

Reemtsma T., Zywicki, B., Stueber, M., Kloepfer, A., Ekel, M. (2002). Removal of sulphur organic polar micropollutants in a membrane bioreactor treating industrial wastewater. *Environ. Sci. Technol.*, 36(5), 1102–1106.

Rosenberger, S., Kubin, K., Kraume, M. (2002). Rheology of activated sludge in membrane bioreactors. *Engineering Life Sciences*, 2(9), 269–275.

Seyhi, B., Droguil, P., Buelna, G., Blais, J.-F., Heran, M. (2011). État actuel des connaissances des procédés de bioréacteur à membrane pour le traitement et la réutilisation des eaux usées industrielles et urbaines. *Revue des sciences de l'eau*, 24(3), 283–310.

Sorensen, H.B. and Jorgensen, E.S. (1993). *The Removal of Nitrogen Compounds from Wastewater*. Elsevier Science, Amsterdam.

Stephenson, T., Judd, S., Jefferson, B., Brindle, K. (2000). *Membrane Bioreactors for Wastewater Treatment*. IWA Publishing, London.

Tao, G., Kekre, K., Wei, Z., Lee, T.C., Viswanath, B., Seah, L. (2005). Membrane bioreactors for water reclamation. *Water Sci. Technol.*, 51, 431–440.

Tardieu, E., Grasmick, A., Geaugey, V., Manem, J. (1999). Influence of hydrodynamics on fouling velocity in a recirculated MBR for wastewater treatment. *J. Membr. Sci.*, 156(1), 131–140.

Van Bentem, A.G.N., Petri, C.P., Schyns, P.F.T., Van der Roest, H.F. (2007). Membrane bioreactors. Operation and results of an MBR wastewater treatment plant. STOWA Report, IWA, London.

Vera, L., Villaroel-Lopez, R., Delgado, S., Elmaleh, S. (1997). Crossflow microfiltration of biologically treated wastewater. *Desalination*, 114(1), 65–75.

Vera, L., Villaroel-Lopez, R., Delgado, S., Elmaleh, S. (2000). Enhancing microfiltration through an inorganic tubular membrane by gas sparging. *J. Membr. Sci.*, 165(1), 47–57.

Vyas, H.K., Bennett, R.J., Marshall, A.D. (2002). Performance of cross flow microfiltration during constant transmembrane pressure and constant flux operations. *Intern. Dairy J.*, 12(5), 473–479.

Wang, Z., Chu, J., Song, Y., Zhang, H., Zhao, X, Li, Z., Yao, J. (2009). Influence of operating conditions on the efficiency of wastewater treatment in membrane bioreactors. *Desalination*, 245(1–3), 73–81.

Wintgens, T., Gallenkemper, M., Melin, T. (2002). Endocrine disrupter removal from wastewater using membrane bioreactor and nanofiltration technology. *Desalination*, 146(1–3), 387–391.

Wintgens, T., Rosen, J., Melin, T., Brepols, C., Drensla, K., Engelhardt, N. (2003). Modelling of a membrane bioreactor system for municipal wastewater treatment. *J. Membr. Sci.*, 216(1–2), 55–65.

Xing, C.H., Tardieu, E., Qian, Y., Wen, X.H. (2000). Ultrafiltration membrane bioreactor for urban wastewater reclamation. *J. Membr. Sci.*, 177(1–2), 73–82.

Yamamoto, K., Hiasa, M., Mahmood, T., Matsuo, T. (1989). Direct solid–liquid separation using hollow fiber membrane in an activated sludge aeration tank. *Water Sci. Technol.*, 21(4–5), 43–54.

Yang, W., Cicek, N., Ilg, J. (2006). State-of-the-art of membrane bioreactors: Worldwide research and commercial applications in North America. *J. Membr. Sci.*, 270(1–2), 201–211.

Yu, K., Wen, X., Bu, Q., Xia, H. (2003). Critical flux enhancements with air sparging in axial hollow fibers crossflow microfiltration of biological treated wastewater. *J. Membr. Sci.*, 224(1–2), 69–79.

Zhang, S.Y., van Houten, R., Eikelboom, D.H., Jiang, Z.C., Fan, Y.B., Wang, J.S. (2002). Determination and discussion hydraulic retention time in membrane bioreactor system. *J. Environ. Sci. Chn.*, 14(4), 501–519.

Zuehlke, S., Duennbier, U., Lesjean, R., Buisson, H. (2006). Long-term comparison of trace organics removal performances between conventional and membrane activated sludge processes. *Water Environ. Res.*, 78(1–2), 2480–2486.

2

Hybrid Systems

2.1. Introduction

Biological processes for wastewater treatment are divided into two large categories: free cultures in solution and cultures fixed on supports. In free cultures, microorganisms are suspended in wastewaters as individual cells or in the form of bacterial flocs. They nourish themselves from the surrounding medium (wastewaters). Fixed cultures are masses of microorganisms attached to inert surfaces, and they are part of a biofilm that becomes responsible for the degradation of wastewater pollutants.

There are two types of free-culture reactors:

1) completely mixed without sludge recycling (ponds and lagoons);

2) completely mixed with sludge recycling.

The activated sludge process is a system of free cultures that draws its name from the fact that settled sludge is still composed of living and/or active sludge. A large part of the bacterial mass is returned to the biological tank to increase the available biomass and thus accelerate biological reactions.

The fixed culture process is a system in which wastewaters are in contact with microorganisms fixed on inert surfaces (bacteria beds, rotating biological contactors, biofilters and fluidized beds). The microbial film is known as biofilm. A biofilm refers to a group of microorganisms and their extracellular products that are attached to a solid surface. Biological organisms that develop on the surfaces originate in the same groups as those of activated sludge systems. Most microorganisms are heterotrophic, with facultative bacteria being predominant. Fungi and protozoa also abound, and algae are present when lightness is available. Species of rotifers, worms, snails, etc. are also observed.

A biofilm consists of a dense and growing layer of microorganisms maintained by a matrix of extracellular polymeric substances (EPSs) on a support surface known as media support. It consists of several layers in which various biological conversion processes occur; their thickness varies when conditions change. The rate of detachment of the substrate by a biofilm depends particularly on its concentration in the liquid phase. At low substrate concentration, its detachment rate will be limited by its concentration, while at high concentration, it will be limited by oxygen diffusion.

An appropriate level of turbulence in the reactor supports the transportation of substrates towards the biofilm and the preservation of its low thickness by shearing. This also facilitates a more uniform distribution of the biofilm on the support surface and a better penetration of substrates. Microorganisms attach and develop in the form of a viscous and slimy substance. Wastewaters containing dissolved organic materials get to the biofilm due to concentration gradients inside the film. Suspended particles and colloids can be retained at the "sticky" surface where they are decomposed into soluble products.

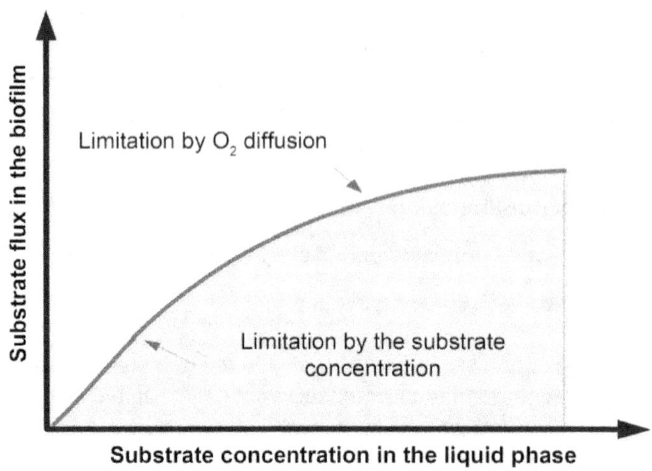

Figure 2.1. *Schematic representation of the effect of substrate concentration on the reaction rate (Morgenroth 2000)*

Oxygen also diffuses in the film. Biodegradation residues diffuse outside the biofilm in the surrounding liquid medium. Therefore, the diffusion process plays an important role in biofilms. Figure 2.1 partially summarizes the various transportation and transformation mechanisms involved in the biofilms. The biofilm grows in a single direction: from the solid surface outwards. Bacterial metabolisms unfold at the biofilm–surface interface (FNDAE 2007). When biofilm grows, the fixation mechanism weakens and the shearing action of wastewater that sweeps the biofilm removes a part of its support. This process is known as sloughing; this designates the loss of surface layers. It is a function of hydraulic and organic loads. The kinetic theory of biofilms relies on the physical fact that substrates that are transported into and out of a biofilm require a concentration gradient for transportation to the biofilm. If the biofilm is sufficiently thick, the pollutant concentration may very strongly decrease in the biofilm and the pollutant will become limiting (Figures 2.2 and 2.3).

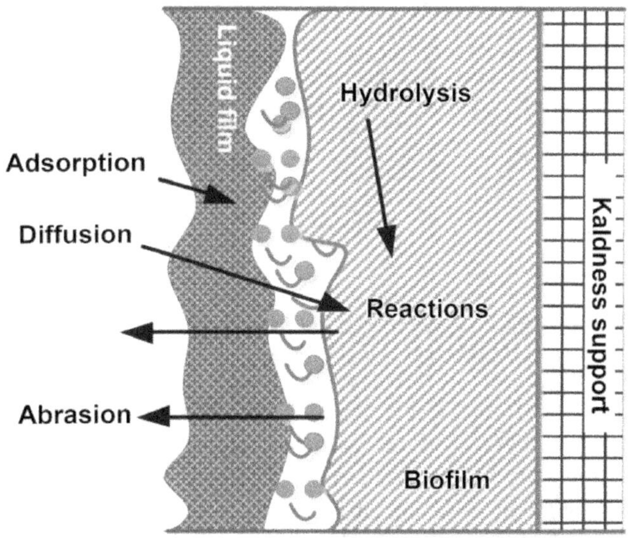

Figure 2.2. *Schematic representation of substrate transportation and of its degradation in the biofilm*

In these systems, the pollutant removal rate depends on many factors such as the wastewater flow rate, the organic rate, pollutant diffusion and that of oxygen in the biofilm and finally temperature. Pollutants and oxygen penetrate deeper when organic load increases.

Hybrid processes refer to reactors integrating bacteria development fixed on a mobile support or fluidized in a free culture of activated sludge. These processes are particularly used to maintain a nitrifying biomass on the support layers. Two processes share a significant part of the market: AnoxKaldnes™ (subsidiary of Veolia Water Technology) MBBR (Moving Bed Bio-film Reactor) and Meteor™ IFAS (Integrated Fluidized Activated Sludge) from Degrémont (subsidiary of Suez France).

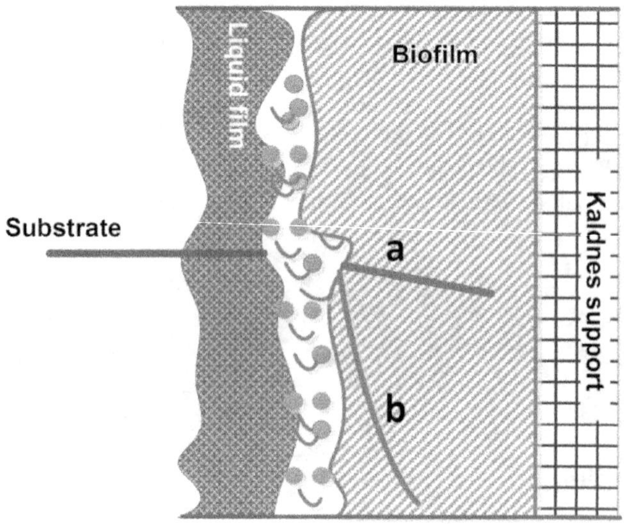

Figure 2.3. *Degradation process: a) metabolism; b) limited substrate*

2.2. MBBR

The AnoxKaldnes™ Moving Bed Biofilm Reactor (MBBR) process is a fixed culture type of biological treatment. It is based on the principle of biofilm with an active biomass developing on specially designed small plastic elements and maintained in suspension in biological reactors. It is therefore a process in which the biofilm develops on supports suspended in wastewater, and it offers a technical solution to remove pollutants. It can be used in nearly all sectors, both in the municipal and industrial markets.

The AnoxKaldnes™ MBBR process uses the advantages of both free and fixed cultures. The reactor is defined as a process with activated sludge that develops throughout the biological tank completed by a mass of microorganisms fixed on a mobile support.

The reactor can be used for aerobic and anoxic processes. It is designed to ensure optimal conditions for the free culture (geometry, aeration, stirring) and follow through biological supports providing a wide active surface for the biofilm. Supports vary in shape and size, which offers the choice to use the best-fitted one depending on the wastewater characteristics, pretreatment, effluent standards and available volumes.

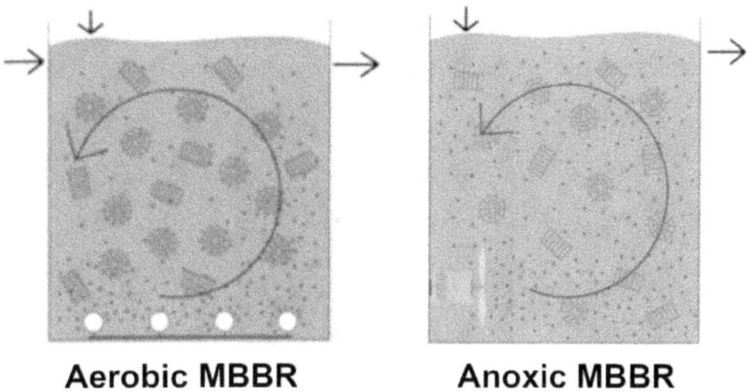

Figure 2.4. *Internal configuration of the MBBR*

Biofilm is very rapidly formed under favorable conditions (aeration, nutrient intake, temperature). Although a biofilm is often visible only after several days, a microscopic layer develops very rapidly inside after immersion in water. After three to four hours, groups of fixed bacteria appear and form microcolonies.

The volume load rate in the reactor (kg $DBO_5.m^{-3}$) influences the biofilm. Very high loads may lead to fungal biofilms (which is also the case under acid pH conditions). Moreover, a high load facilitates the formation of compact biofilms with a small protozoa population. Moderate loads yield a thin biofilm with a rich variety of ciliate protozoa, while a low load may yield very heterogeneous biofilms dominated by protozoa and metazoa.

Studies on the colonization of AnoxKaldnesTM supports have shown that among the first colonizers, pedunculate ciliates (mainly *Vorticella* spp.) play a significant role. When the biofilm further develops, they are replaced by colonial pedunculate ciliates (particularly *Carchesium* spp.). The main food source for most protozoa and metazoa present in wastewater and in the treatment systems consists of enteric bacteria in suspension and not bacteria composing the biofilm. They cannot feed on the biofilm, as they do not have the organs required to feed on bacterial glycocalyx (polysaccharide).

Support materials such as polyethylene, polypropylene and polyvinyl chloride facilitate biofilm formation and its microbial diversity. The functional amine group (-NH2) in the polyethylene and polypropylene increases biofilm thickness and density (Chattopadhyaya et al. 2022).

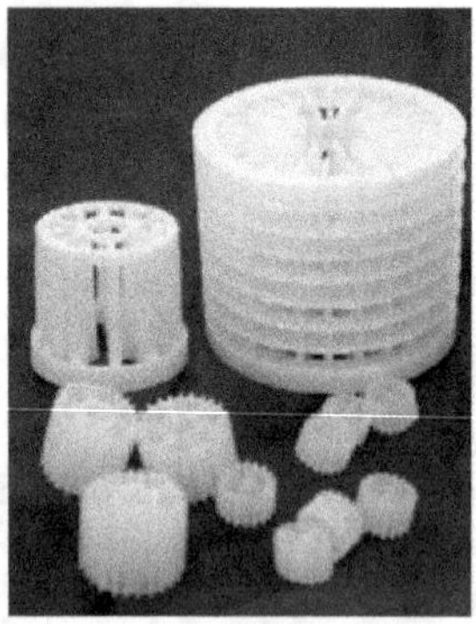

Figure 2.5. *K1 and K2 Natrix supports*

Due to the importance of diffusion, the efficient biofilm thickness (depth of the biofilm in which substrates have penetrated) is an important parameter. Substrate diffusion is normally below 100 μm, as the ideal mobile bed biofilm must be thin and uniformly distributed on the support surface. To reach this, turbulence in the reactor must be controlled, both for the transportation of substrates to the biofilm and for maintaining a low thickness of the biofilm by shearing forces.

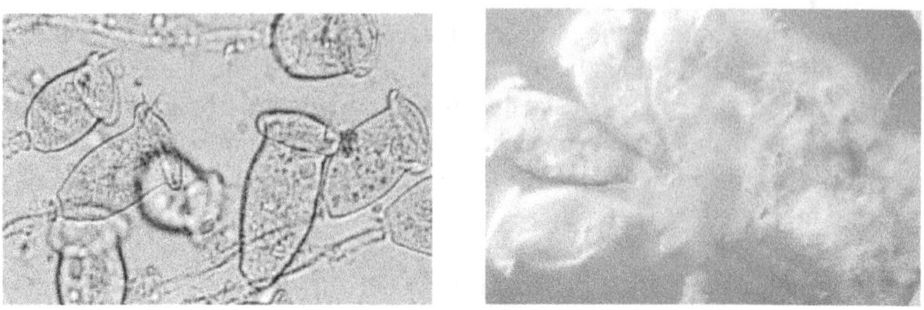

Figure 2.6. *Vorticella*

The main difference between the fixed culture (biofilm) and the free culture (activated sludge) is that free growth tends to contain less anaerobic microorganisms. This is due to the fact there is a lower tendency towards long-term development of anaerobic conditions in the biofilm. The massive transfer of substrates to the biofilm can be far less significant in free cultures, where diffusion is multidirectional, while in fixed culture motion, it is unidirectional.

A significant advantage of the process is that the filling fraction of the supports in the reactor is selected depending on the operating conditions and on the effluent standards. To ensure an optimum motion of supports, the maximal filling fraction should be 67% for aerated reactors and 60% for anoxic reactors. Lower rates are possible in the case of rehabilitation of the existing stations, for which the equipment and operating conditions should be adapted to the geometry and to the work volume. Activated sludge processes refer to volatile suspended solids (VSSs), while for biofilm processes, this parameter is more complex, because it is more difficult to estimate the quantity of fixed bacteria.

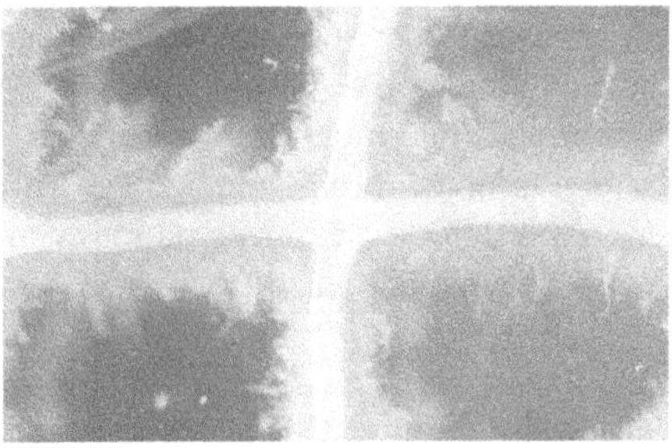

Figure 2.7. *Biofilm rich in pedunculate ciliate colonies*

The average concentration of the fixed biomass is estimated at about 3–4 kg SS.m^{-3} of tank, which is approximately identical in activated sludge reactors.

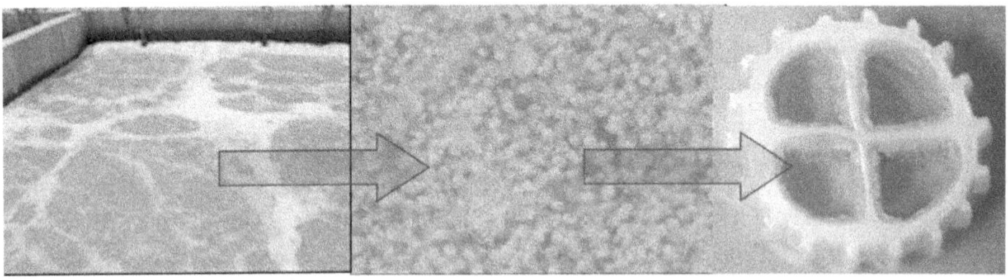

Figure 2.8. *MBBR and biofilm growth*

As shown in Figures 2.7 and 2.8, biomass grows less outside the supports than inside them. This is probably caused by the erosion that occurs when supports collide with each other. This justifies that the effective surface of the supports is estimated at only 70% of the total surface.

As described in Chapter 2 of the first volume of this book, organic material is present in wastewater as proteins, carbohydrates, greases, volatile fatty acids and low or non-biodegradable substances.

Depending on the size of the organic material, organic content can be dissolved, take the form of colloidal matter and be suspended solids. Suspended solids are retained by a filter with pores of about 1 mm. The dissolved organic material can be defined by the fraction containing particles whose size is below 0.1 mm and the colloidal fraction will range between 0.1 and 1.0 mm. Suspended solids consist of an inert fraction of biomass and a slowly degradable fraction.

Colloidal and dissolved materials consist of an inert and degradable fraction and of a slow-mix and readily degradable fraction, respectively. The organic material that cannot be directly used by microorganisms must be hydrolyzed. It is an extracellular degradation of enzymes in dissolved organic material. The inert fraction of the dissolved material will cross through the treatment station without modifying its state. This means there will always be a certain quantity of chemical oxygen demand (COD) in the effluent.

At the end of the wastewater treatment, it is normal to characterize the organic material by measuring the biochemical oxygen demand (BOD) and the COD.

In the aerobic treatment processes, the heterotrophic organisms will oxidize the organic material to form carbon dioxide (CO_2) and water (H_2O). The organic content is also used for the production of the new biomass through the mechanism of bacterial growth.

Environmental factors such as temperature, oxygen availability, pH, nutrients and toxic substances impact the growth of aerobic heterotrophic microorganisms.

Integrated into a wastewater treatment system, the biological treatment process can be presented in the form of an aerated and/or anoxic reactor. In the case of a nitrification–denitrification process, some reactors can be equipped both with a mechanical mixer and an aeration system. This type of reactor will be in anoxic mode during summer, in order to maximize pre-denitrification (to save an external source of carbon). In winter, it will be aerobic to ensure a sufficient aerobic volume for maintaining nitrification at low temperature.

The efficiency of the MBBR process relies to the same extent on the free biological mass circulating in the tank and on that hanging on the supports. The surface colonized by bacteria consists of a solid biofilm that brings a net contribution to the treatment. Reactors can operate under aerobic conditions to eliminate BOD and ammonia or under anoxic conditions for denitrification.

During operation, supports are maintained under constant circulation. In an aerobic reactor, circulation is supported by the action of large and average air bubbles from diffusers. In an anoxic reactor, an immersed mixer is generally provided. Supports can occupy up to 70% of the volume of the reactor. As shown by experience, the mixture is more complicated, and efficiency decreases if the filling percentage is too high.

The treatment capacity of MBBR is a function of the specific surface area (S spec) of the materials. It is calculated as the ratio between the total surface available for the biofilm formation and the volume of the reactor. It is therefore the surface available for the development of the biofilm per unit volume of the reactor. The total specific surface area (TSSpec) of the reactor is equal to the S spec of the medium multiplied by the fraction of the volume occupied by the supports in the biological tank.

2.2.1. *Treatment system*

The treatment system generally includes pretreatment (screening, grit removal, oil removal and degreasing). Then, there is a primary stage whose implementation can be motivated by the following reasons:

– Addresses a problem of load variation: the implementation of a primary stage, with or without reagent, allows the decrease in the coefficient of variation of loads applied to the biological reactor for carbon treatment, provided that it only starts when the load increases.

– Limits the volumes of the biological reactor while maintaining higher sludge production: implementation of a primary settling tank may prove interesting in the case of sludge digestion retained on the site.

– Treats phosphorus by injection of physico-chemical reagents: it is important to note that this treatment can also be implemented at the outlet of the biological reactor, which makes it possible to do without the tertiary treatment. It should also be recalled that the injection of metallic salts for coagulation ($FeCl_3$) gives rise, due to its acidity, to consumption of HCO_3^-, whose lack may impact the downstream nitrification stage, which is detrimental for certain installations located in regions where wastewaters have a low HCO_3^- load.

The treatment can then consist of various reactors in series according to the desired treatment. The fixed biomass is specific to each reactor, as the support does not travel from one reactor to another. The interest of the fixed culture is to maintain a sludge age that is longer than the hydraulic retention time (HRT) of the work, considering the absence of sludge recirculation at the reactor head. During the treatment, the biodegradable soluble pollution is treated, depending on the implemented conditions (concentration of free biomass, sludge age, filling rate in support materials, applied load, reaction time). The non-assimilated biodegradable particulate pollution is trapped by the downstream separation work (clarifier, flotation agent, screen) at the same time as the excess biomass.

The excess fixed biomass comes off and is then trapped in the biomass/treated effluent separation work located downstream (clarifier, flotation agent or mechanical filtration).

Good fluidization is required for the treatment. It depends on various factors such as the type of support, the filling rate of supports in the tanks, the configuration of works, the flow rate of injected air and the air diffusion conditions at the level of the invert or the mechanical stirring power for denitrifying reactors.

MBBRs do not require sludge recirculation at the head of the biological reactor.

Besides the classical treatment parameters (substrate, oxygen, mineral carbon, ammonia nitrogen and orthophosphates), the biological activity will depend on a good contact between the biomass and the effluent to be treated; hence, the necessity of a very good homogenization (fluidization) of the support in the reactor by aeration or stirring. This fluidization is necessary to the treatment to facilitate the substrate/biomass exchanges. The frictions between supports also allow a better control of the thickness of the biofilm, which will contribute to the efficiency of exchanges within the biofilm.

Downstream of the biological stage, the separation of the produced biomass from the treated effluent can be achieved by classical or lamellar settling, by dissolved air flotation or by mechanical filtration by rotary screens.

In the case of classical or lamellar settling, measures must be taken to avoid the "wild" denitrification in the work.

2.2.1.1. *Main configurations*

An MBBR may consist of a single stage, but its staging in several reactors improves the overall effectiveness of the treatment. Reactors facilitate, first of all, the removal of organic material, then the nitrification by minimizing the growth of heterotrophic biomass in those that are downstream.

Standard layout configurations, using this process for oxidizing organic material, nitrification, removal of total nitrogen by denitrification and chemical phosphate removal, are presented in Figures 2.9 and 2.10. Other configurations are also possible. The number of reactors in series varies according to the characteristics of wastewaters and the treatment objectives. To ensure an optimal solid–liquid separation downstream RBGS, coagulant and polymer dosage equipment must be considered in all cases.

The MBBR can be used in various applications:

– carbon removal;

– nitrogen removal;

– industrial wastewater;

– rehabilitation of an overloaded existing waste treatment plant, with two possibilities of configuration:

- implementation of the MBBR process at the head of the system (upstream of activated sludge) has a double objective: maintaining the existing works and reducing the load applied to the activated sludge,

- implementation of the support inside the biological reactor of the existing activated sludge (known as a hybrid system or Integrated Fixed-Film Activated Sludge (IFAS)). This configuration requires a certain number of layouts of the biological reactor, such as the implementation of grids at the outlet and modifications at the level of the distribution of air injected on the invert of the biological reactor. This IFAS process can also be implemented for a new installation;

– new stations with the creation of a complete row equipped with this technology, which means the implementation of several reactors in series depending on the type of treatment in question (pre-denitrification reactor, carbon reactor, nitrification reactor, post-denitrification reactor).

This technology can be used to treat carbon or nitrogen pollution (only nitrification or nitrification–denitrification, in pre- or post- for denitrification) depending on the required levels of discharge. For this purpose, a reactor is installed for each treatment.

In contrast, phosphorus is only treated chemically (addition of metallic salts), at the level of the primary stage (when it is present) or immediately upstream of the separator.

Several MBBR treatment rows are generally installed in parallel, in order to reduce the volume of the biological reactor to facilitate fluidization or to address the problem of load variation (one or several rows are fed depending on the period of the year).

Upstream of the biological stage, the standard treatment system is always composed of a high-performance pretreatment stage consisting of a screening whose mesh does not exceed 3 mm and a grit chamber-oil remover. The implementation of a primary treatment is not compulsory.

In any case, considering the absence of sludge recirculation (except in the IFAS, where sludge is recirculated), MLSS concentrations in the reactor and entering the separation unit are very low (concentrations below 300 mg of MLSS.L^{-1} for domestic wastewater in dry time, a concentration function of the BOD_5 concentration and the MLSS of inlet waters and of the presence or absence of a primary treatment).

The addition of flocculants and/or coagulants is highly recommended for clarification and compulsory for flotation. In this case, it is not systematic for the fabric filtration separation and will depend on the intended outlet MLSS concentration. In this latter case, an addition of $FeCl_3$ at the head of filtration is applied.

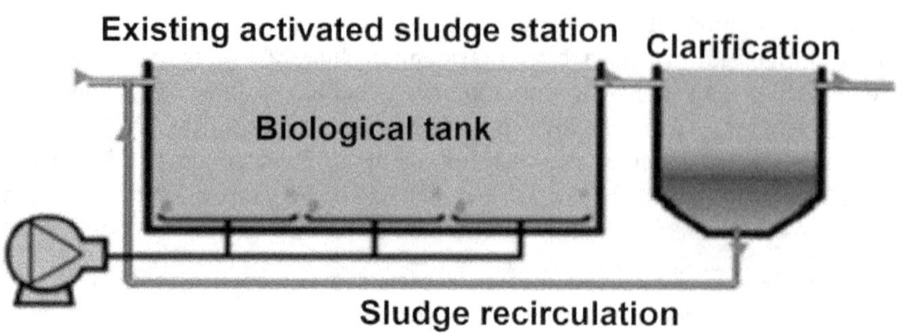

Figure 2.9. *Configuration of the standard activated sludge process*

In a treatment system, an MBBR can be placed in the pretreatment, main treatment and post-treatment. Besides nitrification, an MBBR can be designed to remove nutrients such as total nitrogen and phosphorus. The removal of total nitrogen by denitrification is conducted with a stirred but not aerated MBBR.

The reactor can be installed upstream of other (pre-anoxic) bioprocesses or downstream (post-anoxic) with the possible addition of external organic matter. For pre-anoxic, one part of the nitrified effluent is returned to the head in order to use the oxygen issued from nitrates to oxidize the organic matter from the affluent.

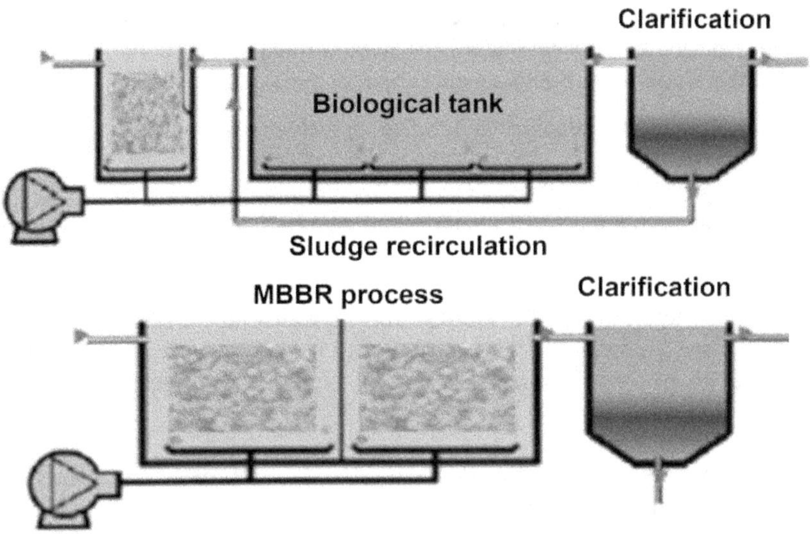

Figure 2.10. *Various MBBR configurations*

Phosphorus is normally removed by coprecipitation. The chemical phosphate is preferably removed by adding coagulant after the MBBR upstream of the secondary clarification. At this stage, the addition of coagulant increases the solid–liquid separation and prevents generating a lack of phosphorus in the biofilm or an excessive reduction of alkalinity. The removal of trace organic contaminants can be better achieved with a biofilm system than with a system with suspension culture. Indeed, the presence of a biofilm of a certain thickness facilitates the appearance and retention of specialized bacteria at the basis of the biofilm (increased sludge retention time). Moreover, the protection offered by the thickness of the biofilm ensures a better resistance to toxic shocks than a free culture process.

2.2.2. *Materials*

In order to obtain a high-performance process with biofilm and optimize the diffusion of substrates, the thickness of the boundary layer and of the biofilm can be controlled by the type of support and its quantity, the wastewater supply system and the hydrodynamic conditions. The thickness of the biofilm typically varies between 50 and 500 μm. Thick biofilms (500 μm) lead to better removal rates of trace organic contaminants difficult to biodegrade, while thin biofilms (50 μm) can be used to obtain higher nitrification rates. There are four stages of biofilm formation, as follows: attachment, accumulation, regeneration and maturation. As the biofilm grows, there is a continuous detachment of the biomass caused by:

– abrasion related to collisions and to friction between supports;

– erosion related to shearing forces of the liquid around the biofilm;

– desquamation related to the detachment of large segments of biofilm of the support.

Given this detachment, an MBBR still contains a portion of biomass in suspension in the liquid phase that contributes to its effectiveness. This contribution depends, particularly, on the rate of organic load, on the substrate concentration and on the HRT.

Table 2.1 presents the specific surface areas of several materials used as biofilm support. MBBR materials have higher values compared to usually encountered classical materials and types of MBBR media. The MBBR has higher potential performances than the standard processes of the fixed film type. MBBRs use screens to separate biological mass agglomerate-seeded supports. The retention of media inside the reactor facilitates the selection and the enrichment for the best-adapted microorganisms that prevail in this steady-state reactor.

	Specific surface area ($m^2.m^{-3}$)
Bacteria beds:	
– Plastic material	45–60
– Rock material	69–150
Biological disks	100–150
Biofilter materials:	
– Biodagene	800–1,200
– Biostyr	800–1,200
– Biolite	800–1,200
AnoxKaldnes™ supports	200–1,200

Table 2.1. *Specific surface areas of several materials used for fixed cultures*

Name	Supports	Size in mm (length x diameter)	Composition	Effective specific surface area ($m^2.m^{-3}$)	Maximum filling rate (%)	D ($kg.m^{-3}$)
K1 (Kaldnes)	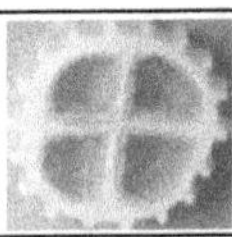	7.2 x 9.1	PE	500	67	150
K2 (Kaldnes)	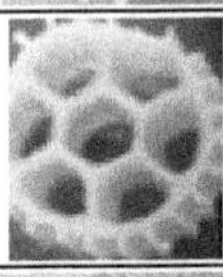	15 x 15	PE	350	67	–
K3 (Kaldnes)	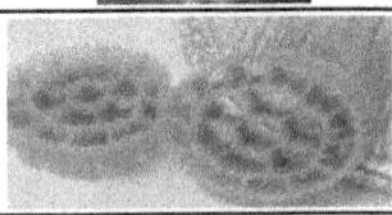	10 x 25	PE	500	67	100
K5 (Kaldnes)	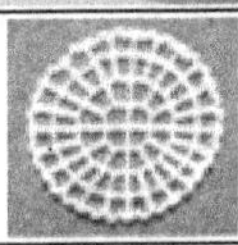	4 x 25	PEHD	800	–	–
Biochip M (Kaldnes)	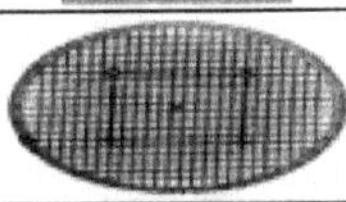	2.2 x 48	PE	1 200	55–60	234
Biochip P (Kaldnes)	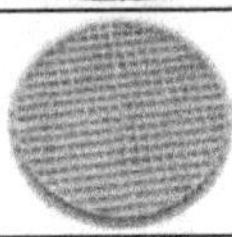	3 x 45	PE	900	55–60	173
Natrix C2 (Kaldnes)	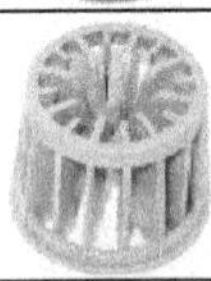	30 x 36	PE	220	65	113
Natrix M2 (Kaldnes)		50 x 64	PE	200	65	80
Z-MBBR (Kaldnes) 50, 200, 300, 400, 500 µm		1.5 x 1.5 2.3 x 2.3	–	1,270–2,420	–	–
Bioports 600-14		14 x 18	–	500	–	–

Name	Supports	Size in mm (length x diameter)	Composition	Effective specific surface area ($m^2.m^{-3}$)	Maximum filling rate (%)	D ($kg.m^{-3}$)
Bioports 900-09	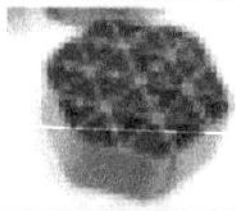	9 x 17	–	942	–	–
Peenox		17.5 x 23	–	400	–	–
Peenox mini		10–15	–	750	–	–
NutriCell DTi1 (ActiveCell)	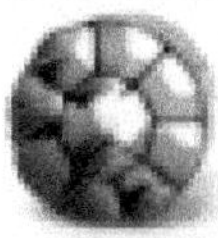	16 x 22	–	402–450	–	–
NutriCell DTi2 (ActiveCell)		16 x 22	–	485–515	–	–
NutriCell DTi3 (ActiveCell)	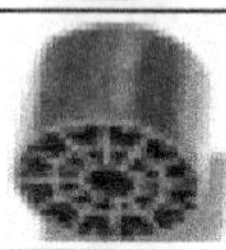	16 x 22	–	535–630	–	–
NutriCell DTi4 (ActiveCell)	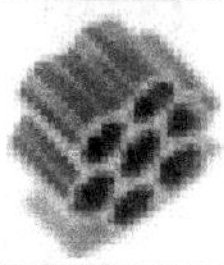	16 x 22	–	637–750	–	–
BWT15 (Vinci)		12 x 15	–	830	–	–
BWTX (Vinci)		15 x 15	–	640	–	–

Table 2.2. *Physical characteristics of several supports on the market*

The two BiofilmChips™ are often preferred for their large specific surface area (900 and 1,200 $m^2.m^{-3}$ ordered against 500 $m^2.m^{-3}$ ordered for the model K1), allowing a significant decrease in the tank volumes. According to Cemagref, this type of media requires a significant operational rigor due to very small interstices presenting the highest risks of clogging. To limit this risk, the provider of this type of support recommends their use only downstream of a primary settling tank or a screen whose mesh does not exceed 1 mm. These recommendations are not always respected.

Model K1 has the following advantages:

– Relatively significant interstices are rarely clogged.

– Average bubbles break more easily into smaller bubbles at contact with supports. This leads to better oxygen transfer rates.

– Setting into motion is facilitated by small fins arranged at the external periphery.

Name	Shape	Size in mm (length x diameter)	Composition	Effective specific surface area ($m^2.m^{-3}$)	Maximum filling rate	Applications
BMX1 (Vinci)		7 x 9	PE	545	67 (C, N)	C + N + DN
Meteor 660 (Degrémont)	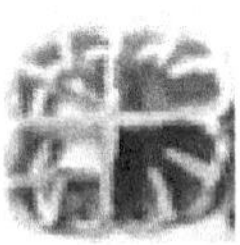	12 x 12.2	PEHD	660	67 (C, N DN)	C + N + DN
Meteor 450 (Degrémont)		15 x 22	PEHD	450	67 (C, N DN)	C + N + DN
Meteor 515 (Degrémont)		15 x 22	PEHD	515	67 (C, N DN)	C + N + DN

Table 2.3. *Physical characteristics of materials of other manufacturers*

As a comparison, BiofilmChip™ products need more energy to be effectively mixed. Mechanical stirring is compatible. Anoxic and anaerobic reactors in which aeration is limited for some and zero for others are stirred by mechanical agitators rather than by injection of air when they are lined with K1 media.

The effective specific surface area corresponds to the surface available for the growth of the biofilm protected from abrasion and shearing per unit volume of media support based on a 100% filling rate. The effective specific surface area is therefore smaller than the total surface area of the support. This parameter is an important criterion for MBBR dimensioning, because it influences the media support volume required in the reactor. The type of support is chosen depending on the targeted applications, in order to keep the biofilm active for the contaminants to be removed. The support used in the aerobic reactors has generally an effective specific surface area of about 500–700 (± 100) m^2 of biofilm/m^3 of media support (Metcalf & Eddy Inc. 2014).

The MBBR filling rate for the supports (%) is determined by calculating the ratio between the volume occupied by the supports and that of the reactor. The percentage of filling by the supports ranges between 25% and 67% of the volume of the reactor. A filling above 70% might prove harmful for the control of the biofilm thickness and to the hydraulics of the reactor (Ødegaard 2000). An initial design at a filling rate of 40–50% allows the potential increase of the capacity of the process by the addition of media support at a higher percentage.

2.2.3. *Dimensioning*

The main parameters for dimensioning MBBR processes are the surface load applied or eliminated per m^2 of useful surface area and per day, associated with efficiency and the associated temperature. The applied surface load has consequences for the obtained efficiency, and therefore the eliminated load and the quality of rejected waters.

The unit used differs depending on the intended type of treatment:

– in g of BOD_5 (or COD) applied/m^2 of useful surface area and per day for the carbon treatment;

– in g of N-NH_4^+ applied/m^{-2} of useful surface area and per day for the treatment of nitrogen in nitrification;

– in g of N-NO_3^- applied/m^{-2} of useful surface area and per day for the treatment of nitrogen in denitrification.

2.2.3.1. *Applied loads*

MBBR process can be used under high, average and low organic loads. An MBBR with average BOD_5 can be used to meet the demands of a secondary treatment, while under a low BOD_5 load, it facilitates simultaneous nitrification in the reactor if there is sufficient aeration. With several works in series, nitrification is generally produced in the reactors located more downstream.

2.2.3.1.1. Surface load

The design surface area loading rate (Rsb in g $BOD_5.m^{-2}.\ d^{-1}$) applied can be used to establish the required volumes for the media support and reactor. It depends on the concentrations of BOD_5 to degrade and of the dissolved oxygen (DO) in the liquid phase of the reactor, the mixture conditions and the characteristics of the biofilm (Metcalf & Eddy Inc. 2014).

It makes it possible to define the total required carrier surface area (Sms, total surface media support) in the reactor as a function of the average flow rate (Qm) of the expected concentration of BOD_5 at the outlet of the reactor (Lf) and of the raw wastewater (Lo):

$$Sms\ (m^2) = Qm\ (Lo - Lf)/Rsb$$

where:

– Qm: daily flow rate ($m^3.d^{-1}$);

– Lo: BOD_5 concentration at MBBR inlet ($mg.L^{-1}$ or $g.m^{-3}$);

– Lf: BOD_5 concentration at MBBR outlet ($mg.L^{-1}$ or $g.m^{-3}$).

2.2.3.2. *BOD_5 removal under high load*

The specific loads applicable to high rates of BOD_5 are presented in Table 2.4. An MBBR with a high organic load allows a partial removal of BOD_5 to increase the capacity of a downstream process and improve the settling properties of activated sludge. As the biological sludge contains about 0.6 mg $BOD_5.mg^{-1}$ MLSS, it is important that the MLSS concentration at the outlet of the system is the lowest possible to obtain a low-level BOD_5.

High loads are applied in the case of:

– pretreatment, upstream a classical/common activated sludge in order to reduce mass loads;

– when a significant reduction of BOD_5 is not required.

Percentage of reduction	**Specific load (g $BOD_5.m^{-2}.d^{-1}$)**	**T (°C)**
60–70	20–25	10
70–80	15–20	10

Table 2.4. *Efficiency of BOD_5 reduction under high load*

2.2.3.3. *BOD_5 removal under average load*

The AnoxKaldnes™ reactor supplied with an average load is used for the final phase of treatment. An MBBR at average BOD_5 load allows meeting the demands of a conventional secondary treatment. Given the BOD_5 content of suspended solids, the MLSS concentration at the outlet must be below 10 – 15 $mg.L^{-1}$ to meet the demand.

Table 2.5 presents the dimensioning.

Reduction percentage	Specific load (g $BOD_5.m^{-2}.d^{-1}$)	T (°C)
80–85	10–15	10
85–90	6–10	10

Table 2.5. *Efficiency of BOD_5 reduction under moderate load*

2.2.3.4. *BOD_5 removal at low load (before nitrification)*

An MBBR at a low BOD_5 load facilitates the nitrification simultaneously to carbon reduction in the same reactor if aeration is sufficient. With several MBBRs in series, nitrification generally occurs in reactors located more downstream.

During nitrogen removal, the volumes of tanks dedicated to BOD_5 and nitrogen are often separated. For the nitrification to take place, all the biodegradable organic matter must be removed. Consequently, for the design of reactors, applied loads are relatively low. Dimensioning is presented in Table 2.6.

Percentage of reduction	Specific load (g $BOD_5.m^{-2}.d^{-1}$)	T (°C)
90–95	2–5	10

Table 2.6. *Efficiency of BOD_5 reduction under low load*

2.2.3.5. *Required volumes*

We refer to indications given in Tables 2.4–2.6 for the specific loads to be applied.

The required volume of supports (Vsp) is calculated by dividing the carrier surface area (Sc, m^2)/carrier-specific surface area required to be protected by the unit volume of media support (Ssp, m^2/m^3).

The calculated required carrier volume will be in m^3.

The volume of the reactor (V_R) is finally determined by dividing this value by the percentage of the MBBR filled by the supports (% sp):

$$V_R = 100\ Vsp\ /\ \%sp$$

with:

– Vsp being the media support volume (m^3);

– V_R being the volume of the reactor (m^3);

– %sp being the percentage of the MBBR filled by the supports.

It is possible to deduce the liquid volume in the tank:

liquid volume in tank = required tank volume – [required carrier volume (1 – carrier % void space)]

In order to convert the surface area loading rate (Rsb) into the volume equivalent (Rsv), the following can be used:

$$Rsv = Rsb.Ssp.\%sp/100{,}000$$

$$Chv = Chs\ Ssp.\%sp/100{,}000$$

with:

– Rsv being the volume loading rate ($kg.m^{-3}.\ d^{-1}$);

– Chv being the rate of the volume load applied on the biofilm ($kg.m^{-3}.d^{-1}$).

2.2.3.6. *Oxygen requirements*

Similar to activated sludge, O_2 requirements include those necessary for metabolism and those necessary for endogenous respiration.

For BOD oxidation, 0.87 kg O_2.kg BOD_5^{-1} are used.

The needs of endogenous respiration are estimated by the following relation:

$$\text{Endogenous respiration} = 2.10^{-3}\ 1.07^{(T-10)}.X_{bf}.V$$

with:

– 2.10^{-3}: O_2 necessary for the biomass (kg O_2 kg^{-1} biomass);

– $1.07^{(T-10)}$: temperature correction (T °C: 5–23°C); if T °C is > 24°C, the relation is simplified to 5.10^{-3} X_{bf}.V;

– X_b: the weight of biomass in the biofilm (4–5 kg.m^{-3});

– V: the total volume of the reactor (m^3).

2.2.3.7. *Sludge production*

The concentration of suspended solids (MLSS) in the effluent of an MBBR depends on the total production of sludge and MLSS of raw wastewater. It may be estimated as the sum of five main components:

1) the heterotrophic biomass produced from the biodegradable carbon substrate;

2) the residues produced by endogenous respiration;

3) the nitrifying biomass produced from the nitrified ammonia nitrogen (one part of the organic nitrogen can be first hydrolyzed into ammonia nitrogen);

4) the non-biodegradable organic matter of the affluent;

5) the inorganic matter of the affluent.

In an MBBR process, the surface concentration of the biofilm varies between 5 and 15 g of MLSS.m^{-2} of support for the simultaneous removal of organic matter and nitrification. It can even reach 28 g of MLSS.m^{-2} for a process with a high load of BOD_5 (Phillips et al. 2008). A volume concentration of MLSS between 3,800 and 5,000 mg.L^{-1} can therefore be reached in an MBBR with a media support having 500 m^2.m^{-3} of specific surface area and a filling rate of 60% (Metcalf & Eddy Inc. 2014). However, the MLSS concentration at the outlet generally ranges between 100 and 250 mg.L^{-1}, which is far below the MLSS concentration of a typical activated sludge process (3,000 to 5,000 mg.L^{-1}).

The quantity of sludge produced can be estimated by considering the sludge production linked to the biomass synthesis, cell debris, of non-biodegradable MVES in the influent and to the hydrolysis of particulate matter (FNDAE 2007; Canler 2011).

In general, the particulate BOD_5 present in raw waters travels more or less unchanged through the reactor. The following relation is used:

$$\Delta X = 0.6\left(\frac{Xo}{Lo} + 1\right) - \frac{0.072 \text{ x } 0.6 \text{ x } 1.072^{(T-15)}}{\frac{1}{G} + (0.08 \text{ x} 1.072^{(T-15)})}$$

with:

– Xo being the SS concentration in the incoming effluent ($mg.L^{-1}$);

– Lo being the BOD_5 concentration in the incoming effluent ($mg.L^{-1}$);

– G being the sludge age (d).

One day of sludge age leads most often to sludge production of 0.4–0.8 kg MLSS.kg BOD_5^{-1}. Unlike an activated sludge process, sludge age is not a commonly used concept for the design and operation of an MBBR. It can nevertheless be estimated based on the rate of removal of MLVSS (mixed liquor volatile suspended solids) from the support (Soleimani Karizmeh 2012). Knowing that the surface layers of the biofilm are more susceptible to detach than those that are near the support, the average sludge age in the liquid phase is quite low (0.5–5 days).

2.2.4. *Operating parameters*

The efficiencies announced by the manufacturers must be systematically accompanied by the applied surface load and the treatment temperature retained in the reactor or at the inlet of the waste treatment plant. There is a one to two degrees Celsius increase in the temperature of raw water at the inlet of the plant and in the temperature of the biological reactor. The total useful surface of the support results from the product of the volume of the concerned reactor and the material filling rate of the reactor and effective surface of the retained support (e.g. the Biochip M support develops 1,200 m^2 of protected surface/m^3 of ordered material). Conversely, the volume of the biological reactor can be calculated from the surface load, BOD_5 flux to treat, the specific surface of the retained support and the filling rate.

The following formulas are used:

– total useful surface of support in m^2 necessary for the treatment = flux of BOD_5 to be treated per day/dimensioning surface load retained for treating BOD_5;

– volume of material required for the treatment in m^3 = total useful surface of support/specific surface of the retained support;

– volume of biological reactor in m^3 to treat BOD_5 = required volume of support/filling rate.

Treatment effectiveness requires an excellent fluidization that strongly depends on the type of support retained and on its filling rate. Each support is characterized by a maximal filling rate (trade-off between compactness and energy powers required for fluidization). A support with a very large specific surface associated with its maximal filling rate will facilitate a gain in compactness and therefore in covered surface. However, compactness is limited by the constraints of the installation of the system of air diffusion to the floor, the passage sections of the hydraulic flux (inlet and outlet of works) and the velocity at which the material is carried by the flow (FNDAE 2007; Canler 2011).

The volume load, corresponding to kg of applied BOD_5 per m^3 of reactor and day, can essentially be used to compare installations in terms of compactness for the biological reactors and provide information on the retention time of effluents in the reactor.

The retention time obtained from the volume load does not take into account the volume occupied by the support. The actual retention time is between the retention time without support and the retention time with support, without considering the water volume inside the support. The retention time with support is obtained by multiplying the retention time without support by the percentage of water in the reactor obtained by the formula 1 – filling rate. In the MBBR process, the soluble organic matter can be degraded by oxidation and microbial synthesis or produced by the hydrolysis of a portion of the particulate organic matter trapped in the irregularities of the biofilm. The resulting particulate matter (produced or not hydrolyzed) must be removed from the effluent of the MBBR by a solid–liquid separation process through which the MLSS targets in the effluent can be reached. If the totality of particulate matter is removed after the reactor, the maximal removal rate reachable depending on the rate of load applied in total COD would be quite high (Figure 2.12).

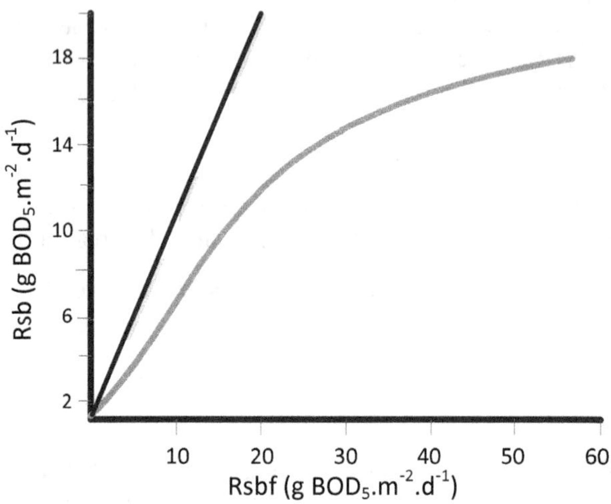

Figure 2.11. *Organic matter removal rate as a function of the applied loading rate in units of BOD_5 filtered at 11°C and 6.2–6.4 mg $O_2.L^{-1}$ with a ratio of 1.6 g raw COD/g BOD_5 (Odegaard 1999; Melun 2004)*

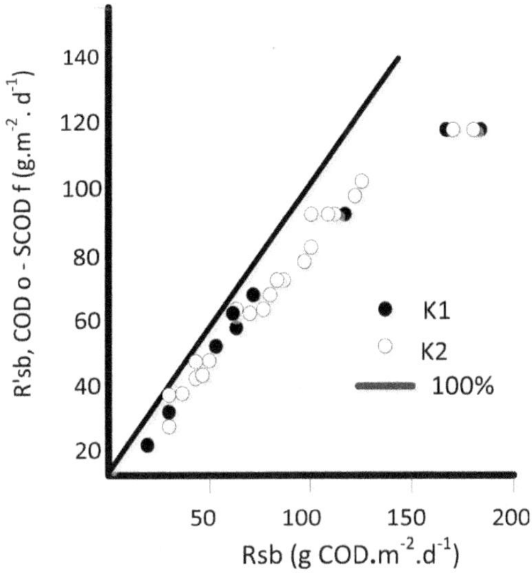

Figure 2.12. *Maximal removal rate reachable as a function of the loading rate applied in total COD with complete removal of particulate materials*

Figure 2.12 presents the organic matter removal rate (R'sb), which is the rate of surface removal by the biofilm (g COD removed.$m^{-2}.d^{-1}$) as a function of the applied loading rate (Rsb), which is the total surface area loading rate, expressed in filtered COD.

According to this figure, R'sb seems to tend towards a maximum of approximately 18 g.$m^{-2}.d^{-1}$ at an Rsb of filtered COD above 45 g COD.$m^{-2}.d^{-1}$. The operating conditions are the following: average T °C of 11°C and concentration of DO of 6.2–6.4 mg of $O_2.L^{-1}$ (McQuarrie and Boltz 2011).

The relation obtained can be used to estimate $R'sb_{COD}$ as a function of Rsb_{COD}:

$$R'sb_{COD} = -0.0064\ Rsb_{COD}^{2} + 0.667\ Rsb_{COD} + 0.311$$

with:

– $R'sb_{COD}$ being the rate of the surface loading rate of removed soluble COD (g.m^{-2}. d^{-1});

– Rsb_{COD} being the total surface loading rate of applied COD (g.m^{-2}. d^{-1}).

2.2.4.1. *Case of carbon A treatment: new installation*

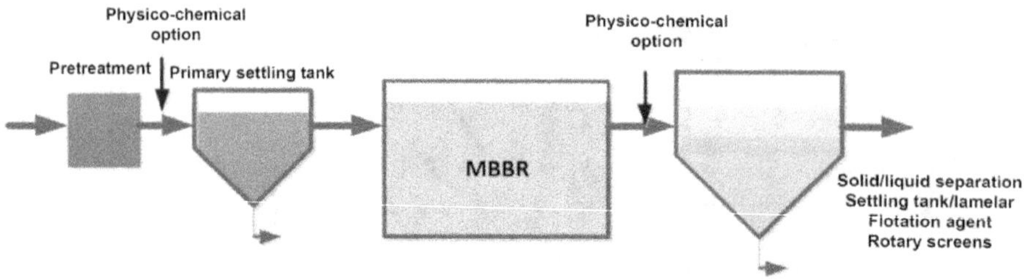

Figure 2.13. *Configuration of an MBBR system for carbon treatment for a new station*

The retained surface loading is a function of the expected efficiencies and of the temperature of the effluents to be treated. In all cases, performances depend also on the retained material, on its degree of fluidization (therefore on its filling rate) and on the absence of a limiting factor (mainly DO and nutrients).

2.2.4.2. *Rehabilitation of an existing plant*

As previously, the retained surface loading is a function of expected efficiencies and of the temperature of effluents to be treated. In all cases, performances depend also on the retained material, its degree of fluidization (therefore its filling rate) and the absence of a limiting factor (oxygen, nutrients).

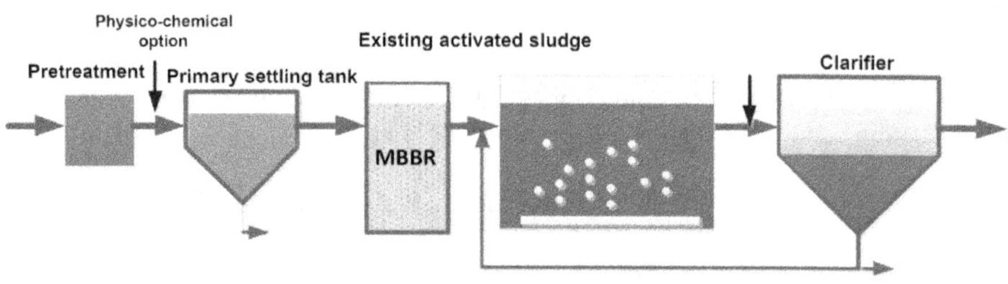

Figure 2.14. *Carbon treatment configuration for a rehabilitation*

It is possible to implement a separator between the MBBR and the activated sludge for better control of sludge ages for each stage of treatment.

2.2.4.2.1. Retention time

The HRT makes it possible to ensure that its value is sufficient for the reaction kinetics to be optimal. In an MBBR, HRT can vary between 30 and 120 min for the removal of BOD_5, depending on the temperature, applied loadings and the quality objectives of the final water. At an average design flow rate, ts generally reaches several hours.

The minimal retention time is obtained by dividing the volume of the reactor by the peak hour flow rate (Q_{max}):

$$HRT = V_R/Q_{max}$$

with:

– HRT being the hydraulic retention time (h);

– V_R being the volume of the reactor (m^3);

– Q_{max} being the peak hour flow rate ($m^3.h^{-1}$).

Too high a value for the HRT intensifies the activity of the suspended biomass to the detriment of the biofilm (Piculell 2016), which risks detaching and increasing the SS concentration in the effluent. Several factors have a great influence on the design. The most important of these factors are mentioned below.

2.2.4.3. *Temperature*

Similar to all biological treatment processes, water temperature influences the activity of biomass in the MBBR. The effect of temperature on the dimensioning of reactors is anticipated by theoretically correcting the applied loading rates on the surface of biofilm with the Arrhenius relation. The effect of temperature on the specific loading rate is expressed by the following relation:

$$r_T = r_{10}\ \Theta^{(T-10)}$$

with:

– r_T being the specific loading (kg $BOD_5.m^{-2}.\ d^{-1}$) at T (°C);

– r_{10} being the specific loading at 10°C;

– Θ being the coefficient of activity related to T °C; Θ is 1.06 for temperatures ranging between 10 and 20°C; Θ is 1.07 for temperatures ranging between 5 and 10°C;

– T is the temperature (°C).

The value of the temperature–activity (Θ) coefficient varies as a function of temperature. Theoretically, the reaction rate will increase up to a maximum at about 36–37°C, but it is recommended not to correct the temperature of the loading rate below 25°C.

2.2.4.4. *DO concentration*

An MBBR reactor intended to eliminate organic matter must be designed considering a DO concentration of 3 $mgO_2.L^{-1}$ in the biological tank. As the organic substrate and the oxygen penetrate into the biofilm, it is important that the DO is sufficient. Indeed, if the oxygen concentration is below 3 $mgO_2.L^{-1}$, oxygen can be consumed before being able to penetrate throughout the thickness of the biofilm. Substrate removal would then be limited and would reduce the effectiveness of the system.

2.2.4.5. *Nutrient elements*

The correct operation of an MBBR requires the availability of nutrients (nitrogen and phosphorus) in sufficient quantities according to the usual ratios: $BOD_5/N/P$ = 100/5/1.

2.2.4.6. *pH*

The pH must also be maintained as stable as possible within a range from 6 to 8. Too low or too high, this parameter may inhibit the process and facilitate the growth of filamentous organisms. A pH adjustment should then be considered.

Nitrogen (ammoniac, nitrates and organic nitrogen) is discussed in Chapter 1 of volume 3.

2.2.5. *Implementation*

2.2.5.1. *Pretreatments*

A pretreatment is required upstream of an MBBR to prevent the accumulation of debris in the reactor (e.g. cloth, plastic, sand) that would be difficult to withdraw in the presence of supports. An average (generally 6–12 mm) screening is recommended when primary settling is planned. In the absence of a settling stage, a thin screening (~ 3 mm) is required upstream of the MBBR in the secondary treatment. The size of this mesh is mainly related to the type of support retained in biological reactors, but it must also take into account the diameter or the spacing of the grid mesh retained at the inlet and outlet of reactors. A grit removal might be necessary, especially in the presence of a unit sewer system. For the grit removal/degreasing, the values of the dimensioning parameters are standard. It is however recommended to make sure that the degreasing station is correctly dimensioned to avoid any introduction of particulate grease in the reactors – these being harmful to the treatment due to the risks of clogging supports and grids and their stagnation on each biological reactor.

The addition of chemical products to primary settling facilitates the improvement of performance and helps to reduce the organic loading intake to the MBBR and contributes to chemical phosphate removal.

However, measures should be taken to maintain a sufficient phosphorus concentration for the subsequent biological treatment. Therefore, coagulant dosages should be limited and the addition of a downstream point of injection is recommended to complete phosphorus removal. This first stage of treatment is essential to all the treatment systems and more specifically to the MBBR process.

2.2.5.2. *Biological work equipment*

The equipment integrated in an MBBR reactor includes:

– a plastic support for the biofilm with a range of available types, as presented in Tables 2.2 and 2.3;

– grids for the retention of supports in the biological tank: the grids, located at the outlet of the reactor, prevent any loss of supports with the treated water. The design of the grids depends on the operating conditions of the tank and on the type of reactor. Most installations are designed with rectangular grids placed vertically for anoxic reactors or with cylindrical grids placed vertically or horizontally for aerobic reactors;

– air diffusers providing average size bubbles: these are dimensioned as a function of the needs of the process and the type of support;

– mixers (anoxic, aerobic reactor): these provide a homogeneous mix of this heterogeneous medium with a biological mass of free bacteria and seeded supports guaranteeing the removal of organic and nitrogen pollutants without damaging the materials;

– antifoaming systems, present in the form of sprinklers or air pumps;

– grids placed at the inlet and outlet of each unit in order to maintain the supports in reactors; their mesh is smaller than the diameter (or the width) of supports.

In the aerobic processes, biofilm supports are maintained in suspension by the stirring achieved by air diffusers, while in anoxic processes, a mixer provides the energy necessary for maintaining the supports in motion.

2.2.5.3. *Aeration*

Besides providing the oxygen required for heterotrophic and nitrifying bacteria, the aeration ensures a uniform distribution of supports maintained in suspension in the liquid phase of the aerobic reactor. The installation of piping and diffusers at the bottom of the reactor must facilitate a spiral motion of the media support and the cleaning of the effluent strainer. In order to prevent dead zones and the deposit of solids, the aeration system must cover most of the bottom of the reactor without however fully covering it (McQuarrie and Boltz 2011; WEF 2011). Vertical pipes equipped with their own air flow control valve are generally installed every one to three lateral pipes to facilitate the control of the spiral motion (WEF 2011; Daigger and Boltz 2018). Aeration is generally conducted with air bubbles of medium or large size (6–12 mm diameter). Thin bubble diffusers (2 mm) have sometimes been used, but they have a higher risk of being clogged by the formation of precipitates or the accumulation of debris (Stenstrom et al. 2008). From a practical point of view, it is preferable to install an aeration system with medium or large bubbles composed of perforated pipes made from stainless steel that are more robust. Due to their larger dimension and to the turbulent air flow through the holes, they require less maintenance and the uniform distribution of media support with a spiral motion can be more readily obtained. Due to their solid structure, the weight of supports can be sustained when the reactor is drained. The detailed characteristics of the aeration system generally used in an aerobic MBBR are summarized by the Water Environment Federation (WEF 2011). The continuous motion of supports generates a shearing effect that improves the efficiency of oxygen transfer of large bubble diffusers. A specific rate of oxygen transfer in clear water from 3% to 3.5% per meter deep is typically observed for MBBRs with medium or large bubble diffusers. The correction factors α and β of these diffusers specific to MBBRs for the calculation of aeration are 0.8 and 0.95, respectively (WEF 2011).

2.2.5.4. *Tank depth*

The aim is generally the highest possible water depth, as the oxygen transfer and the mix are better. The usual depth of water ranges from 5 to 10 meters. A water depth above 10–12 meters should be avoided because of the risk of increasing the CO_2 level in water.

2.2.5.5. *Tank geometry*

The AnoxKaldnes™ process is flexible and can be used to obtain nearly all shapes or dimensions with low or high water depths. Moreover, circular, rectangular, square or triangular reactors (concentric reactors) can be used for the MBBR process.

However, there are different demands for aerobic and anoxic reactors, certain dimensions being more optimal than others.

2.2.5.6. *Circular*

The circular-shaped reactor is the easiest to design, as no geometric considerations need be taken into account. The outlet and inlet screens must be inclined at least at 90°, the opposite being preferred.

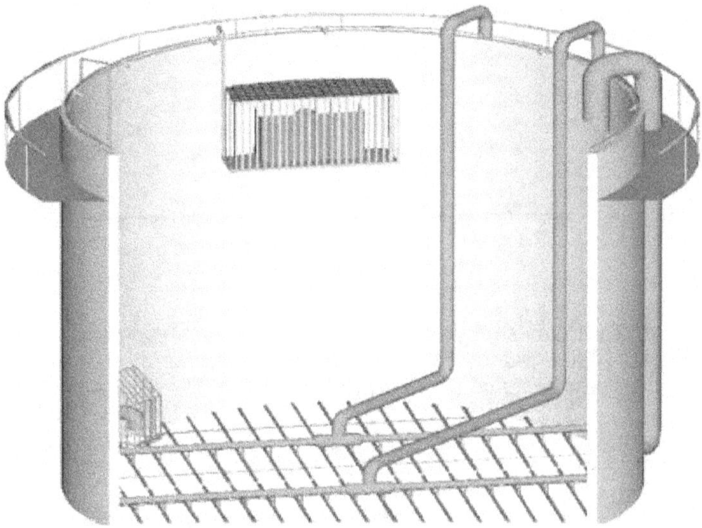

Figure 2.15. *Circular tank (AnoxKaldnes 2011)*

2.2.5.7. *Rectangular*

For the rectangular reactor, it is recommended to avoid long and narrow reactors, which creates a significant hydraulic flux. To design the shape, the value of the water flux must be verified according to Figure 2.16.

This figure shows the acceptable value of the water flow rate in m^3/h per m^2 of cross-section surface as a function of the length/width ratio:

$$\text{Flux } (m.h^{-1}) = \text{flow rate } (m^3.h^{-1})/\text{inlet surface } (m^2)$$

As an example, the flux must be below 30 $m^3.m^2.h^{-1}$ for a ratio L/l = 1.0. It is below 20 $m^3.m^2.h^{-1}$ in the case of a ratio of L/l = 2.25.

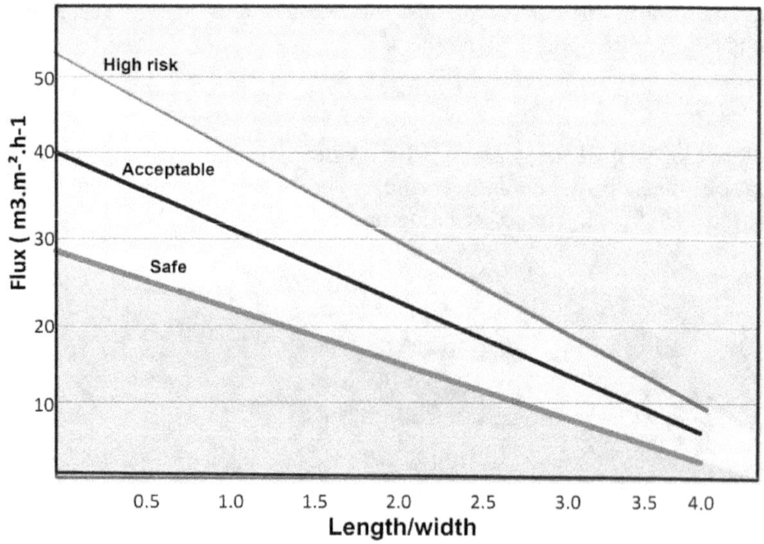

Figure 2.16. *Limits recommended for fluxes supplying rectangular tanks (AnoxKaldnes 2011). For a color version of this figure, seewww.iste.co.uk/gaid/wastewaterengineering2.zip*

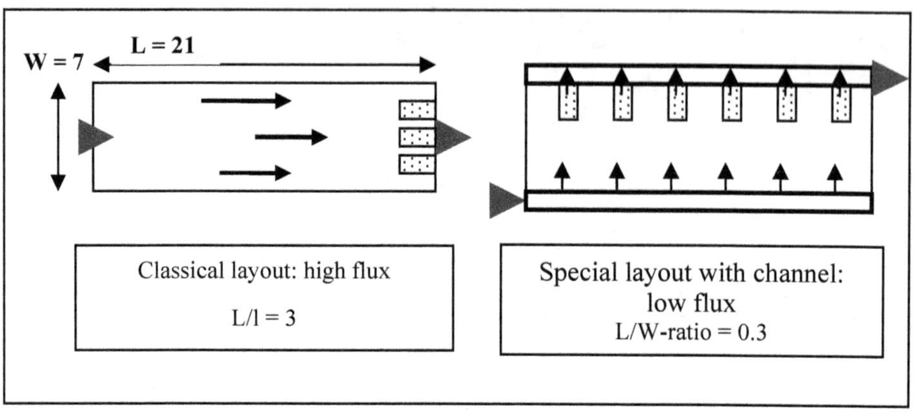

Figure 2.17. *Arrangement of screens as a function of the adopted system (AnoxKaldnes 2011)*

For long/narrow tanks where the hydraulic flux exceeds the limit shown in Figure 2.16, special precautions should be taken:

– Special layouts of channels and outlets using channels and screens of horizontal pipes in the longitudinal direction can provide a good distribution and are recommended (Figures 2.17 and 2.18).

– The layout of the air diffuser should also be taken into consideration, with the possibility to create an upward rotational motion by over-aeration at the outlet of the reactor with diffuser collectors separated in length or in width to give a low rotational flow rate.

Figure 2.18. *Arrangement of screens in a rectangular tank (AnoxKaldnes 2011)*

Figure 2.19. *Material in a reactor (AnoxKaldnes 2011)*

In aerated reactors and for extremely short reactors with a lot of wall effect, flat vertical screens are preferred to horizontal screens "with pipes" to avoid short-circuit flows. Horizontal pipe screens can be considered if their arrangement prevents short-circuit fluxes. A filling degree below 50% is also recommended.

2.2.5.8. *Filling*

There is no minimum filling degree except for the pure MBBR process, for which the filling degree should not be fixed below 35%. But a too low filling degree (< 20%) may cause a bad oxygen transfer in the reactor.

The filling degree is determined as a function of the applied surface loading rate or of the HRT and of the type of fluid.

In industrial applications, removal efficiency is influenced by the HRT and also by the surface loading rate; consequently, the decrease in the volume of reactor and in the HRT may not always be compensated by a higher filling rate.

In municipal applications where wastewaters are readily biodegradable, the HRT is less essential provided that it is longer than 30 minutes. The size of the reactor can therefore be optimized by modifying the filling rate.

Before starting the filling, two conditions must be met:

– The reactor must already be filled with water at a minimum of 70% of the volume.

– The aeration or mixing system must be operational in order to avoid any damage by impact on the carriers. The aeration system is verified in advance by AnoxKaldnes™ in clean water.

The support floats when it is new and dry due to the surface tension resulting from manufacturing and the air bubbles attached to the support. In general, at least two to three days are required for immersion and obtaining a good mix.

Upon starting the operation of an installation, the loading of the supports in the reactors must be conducted according to a specific procedure integrating the know-how of the manufacturer to facilitate the rapid wetting of the support and prevent its stagnation at the surface of tanks. Consequently, fluidization is facilitated by the seeding of the support.

The main characteristics of the aeration system are the delivery of medium/large bubbles and ensuring a complete mix of supports. For these reasons, grids of perforated tubes are recommended and widely used in MBBRs.

Figure 2.20. *Filling a tank with support material (AnoxKaldnes 2011)*

In the choice of boosters, the flow rate range and number are important factors to be taken into account. They will be necessarily equipped with a speed controller and interconnected to allow greater operational flexibility. An emergency apparatus is also compulsory.

Devices should be selected to cope with various operational configurations of the process: a single fed row, or two rows, and a single reactor N or reactors C and N in series. Their flow rate range must be able to supply the air necessary only for fluidization (unsupplied reactor, nighttime period without load to be treated) and the air required for fluidization and for the treatment at full dimensioning load.

Their operation is controlled by measurements of O_2 dissolved in the reactors, with high and low thresholds and various timings (minima and maxima of operational times and time-out).

2.2.5.9. *Mixer in anoxic reactors*

It is important to optimize the mixing intensity in the reactor. A high intensity will reduce the thickness of the boundary layer and of the biofilm, increasing the reaction rates and facilitating a continuous detachment of excess biomass (scouring). However, higher shearing forces on the biofilm and suspended flocs risk causing the slowdown of colloidal particles in the solution, which will increase the sludge volume index (SVI) and decrease the sedimentation rates (Rahimi et al. 2011). The minimal air flow rate required for the uniform distribution of media support in the reactor typically varies between 5 and 10 $m^3.h^{-1}.m^{-2}$ of floor (McQuarrie and Boltz 2011).

The mixer is used in the denitrification reactor to stir the liquid phase and ensure a uniform distribution of supports. The mixing of plastic supports by a mechanical mixer is a very delicate technical point. There is a risk of damaging the support or destroying the mixer, and it is also quite difficult to mix floating supports when no turbulence is applied. In this case, retained supports should be adapted, which means supports with good mechanical strength. As an example, Biochip is not adapted for reactors equipped with agitators.

Submersible mixers with horizontal shafts slightly inclined downwards (15–30°) and on rails have been especially designed for use in MBBRs. The large diameter rotor turns at a relatively low speed (105 rpm at 60 Hz). It generally has three blades whose rounded edges are curved backwards to minimize the abrasion and avoid damaging the media support. The mixing energy to be provided is typically 25 W/m³ of reactor. The ABS SB 1236/7 KA range of mixer/flow boosters was specially designed for MBBR applications. All the parts in contact with the medium (propeller, motor case) are made from Inox 316 Ti. This material has a long term resistance to contact with plastic supports. The propeller has a specific design with blades and the rounded edges are designed to reduce losses by friction and limit damage to the support.

As the media support tends to float, mixers are located fairly near the surface, but far enough to avoid driving the air, which would reduce the denitrification performance (McQuarrie and Boltz 2011).

For a common tank, the power demand per m³ is about 20 to 25 W/m³. For small tanks, it ranges between 30 and 35 W/m³.

Figure 2.21. *ABS SB 1237 KA mixer/flow booster (AnoxKaldnes 2011)*

2.2.5.10. *Removal of treated effluent*

The role of containment strainers and of vertical screens located at the outlet of the tank is to let the effluent pass while containing the supports in the reactor (Figure 2.18). Containment elements such as wire mesh or perforated plates are generally made from stainless steel. Strainers are mainly used in aerobic reactors (BOD_5 removal and nitrification). They usually consist of horizontal cylinders submerged at 35–65% of depth (sometimes 15%) on the outlet wall. Vertical screens mainly used in anoxic reactors (denitrification) consist of flat panels at a distance of 0.15–0.30 m from the outlet wall (open near the surface). MBBR hydraulics is closely linked to the characteristics of media support containment equipment.

To ensure a uniform distribution of supports, a length:width ratio of the reactor (L:l in $m.m^{-1}$) ranging between 0.5 and 1.5 is recommended. Higher ratios (up to 3:1) are possible if the designer proves that the approach velocity is maintained under the recommended limit of 30–35 $m.h^{-1}$. Higher velocities risk driving the media support to the outlet of the reactor and their accumulation on the containment grid would have a negative impact on the flow. The approach speed is calculated as a function of the cross-section of the reactor and of the peak hour flow rate (including recirculation, if that is applicable) coming out of the latter:

$$va = Qmax/(H\ l)$$

where:

– va is the approach velocity ($m.h^{-1}$);

– Qmax is the peak hour flow rate ($m^3.h^{-1}$);

– H is the depth of the reactor (m);

– l is the width of the reactor (m).

The rate of surface hydraulic loading (Chs) through the removal equipment can be calculated by dividing the peak hour flow rate by the immersed surface of the media support containment grids. The rate of hydraulic loading through the total surface of the grid on a wall under peak condition is usually limited at 50–60 $m^3.m^{-2}.h^{-1}$ for a maximal loss of load ranging between 50 and 150 mm (McQuarrie and Boltz 2011; Metcalf & Eddy Inc. 2014). To control this loss of load, a velocity through the holes below 0.5 $m.s^{-1}$ is generally aimed for. If higher hydraulic loads are applied up to 85 $m^3.m^{-2}.h^{-1}$, the approach velocity should consequently be reduced (WEF 2011):

$$Chs = Qmax/S_{grid}$$

where:

– Chs is the rate of the surface hydraulic load ($m^3.m^{-2}.h^{-1}$);

– Qmax is the peak hour flow rate ($m^3.h^{-1}$);

– S_{grid} is the total surface of the media support containment grid (m^2).

For aerobic reactors, with K1, K3, Biofilm Chip M and P supports, a perforated horizontal cylindrical tube is recommended.

For anoxic reactors, with the K1 support, only a vertical plate of perforated mesh is applied in order to avoid disrupting the mixer.

The total surface of the screen is calculated as a function of the peak hour flow rate ($m^3.h^{-1}$), which includes recirculation flow rates, and of the face velocity ($m.h^{-1}$) of each type of screen.

The opening between the meshes of strainers or of screens must be smaller than the smallest dimension of the media support to prevent it from being carried away with the effluent. With a larger-sized media support, it is possible to use outlet grids with larger openings, reducing the load losses through the reactor wall.

Grids must also be installed if there are small openings at the bottom of the walls to allow equalizing the water level between reactors. In the presence of an overflow, the reactor should also be covered by a grid to prevent media support dispersion in the environment.

A series of diffusers with large air bubbles is installed in the containment strainers of the aerobic reactor to facilitate a rotational motion of water and displace a large part of the accumulated material. For anoxic reactors, a single large bubble air diffusion pipe can be installed in front of the vertical screen. Aeration is intermittent; its usual duration is 5 minutes per hour, and the flow rate is about 170 $Nm^3.h^{-1}$ (McQuarrie and Boltz 2011).

2.2.6. *Advantages of MBBRs*

The main advantages of this technology are the following:

– specific to fixed cultures: this means a high tolerance to variations of hydraulic loads and the possibility to rapidly restart an installation after a downtime;

– tolerance to toxic shocks compared to suspended cultures;

– removal efficiency comparable to or higher than that of conventional biological treatment systems with higher rates of volume loads;

– good control of the thickness of biofilm in most applications and easier to clean supports compared to other fixed film technologies;

– flexible design of the reactor (layout, dimension and material), which often facilitates the use of existing structures;

– easy to build and operate;

– rapid start and rapid recovery after stoppage;

– possible progressive increase of the capacity by modifying the filling degree;

– characteristics the process: simplicity of operation (equivalent to activated sludge for the biological reactor part), compactness, a filling rate in supports that can be adjusted as a function of the load to be treated and the absence of washing to eliminate the formed biomass;

– there are many supports with specific characteristics on the market. Each support has its field of application, type of treatment aimed at and maximal filling rate;

– each support is characterized by its specific surface area expressed in m^2/m^3 of material.

At this point, it should be noted that the effective specific surface area commonly retained is the useful surface area corresponding to the protected surface area of the support. Therefore, the external surface area of the support which is subjected to abrasion forces related to supports impacting against each other during fluidization is not available for the fixed biomass and therefore is not integrated in the useful surface.

Similarly, one part of the internal surface of the material may not be accessible to the substrate (as in the case of supports with low porosity resulting in poor circulation of the substrate through the material) or may develop very significant specific surfaces, part of which will not always be accessible.

Caution is also recommended in relation to the material (or support) volume parameter. Depending on the case, this may represent a bulk support volume or a volume of stored and well-arranged material in 1 m^3. The effective surface for a given support refers to the m^2 of protected surface reduced to m^3 of well-arranged material. This data is communicated by the manufacturers, and there are significant differences in effective surface between an ordered material and a bulk material, of about 20 to 30% depending on the retained supports. Some manufacturers provide announced surfaces corresponding to the surface of 1 m^3 of bulk material in order to maintain the consistency with the filling rate. In any case, it should be clearly specified whether the announced surfaces are reduced to m^3 of bulk or well-ordered material.

The best-known supports in terms of performances are the supports developed by AnoxKaldnes™, the company that developed the process.

It should be noted that each support has a maximum filling rate, representing the best compromise between compactness and fluidization.

2.2.7. *Disadvantages of MBBRs*

– Effective screening and pretreatment are required to prevent early clogging of the support.

– Risk of clogging the outlet grids if the process is not correctly operated or designed.

– Energy consumption may be higher than with activated sludge plants given the higher operational level of aeration and the higher contribution of energy in anoxic reactors for the necessities of a mechanical mix.

– Low retention time in the reactor, therefore little hydrolysis of suspended solids. Hence, the biological treatment stage must be followed by an effective separation of particles to eliminate particulate organic materials.

– Continuous renewal of the biofilm because of abrasion, leading to a reduced sludge age in this part of the biofilm. Sludge exiting the reactor with the effluent is not stabilized.

– The aeration system should require no servicing to avoid support withdrawal during maintenance or in case the aeration system cannot be lifted through the support layer.

– Non-stabilized sludge requires additional treatment (e.g. digestion) before being used in agriculture or discharged.

The dimensioning of MBBR processes yields relatively short retention times in the units. Consequently, during a peak (hydraulic and organic) load period, the time of contact between biomass and the pollution to be treated is reduced. In this context, any sudden and significant pollution may temporarily impact the quality of effluent water.

For example, the management of the returns from the sludge dehydration system should be clarified, the objective being to avoid significant and punctual inputs of centrates concentrated at the head of the installation and particularly for nitrogen. In some cases, a possible solution may be a buffer tank with low flow rate reintroduction of these returns in time. Briefly, the data to be specified for a project are the following:

– characteristics of the support used (diameter, thickness and inside spacing of the partition) for each reactor;

– the specific surface area developed per m^3 for each support, specifying the useful or total surface and also if the total surface per m^3 of support is reduced to m^3 of bulk or well-ordered material;

– the volume of each biological reactor dedicated to the treatment;

– the retained and maximal filling rate;

– the removed specific load retained for the treatment of carbon (kg of removed BOD_5/m^2 of useful surface and per day) and of nitrogen (kg of removed $N\text{-}NH_4^+/m^2$ of useful surface and per day).

2.3. IFAS

2.3.1. *Operating principle*

The supports used for MBBRs can also be added to activated sludge reactors, leading to the IFAS process. Integrating a fixed culture in activated sludge may prove particularly interesting when significant nitrification in cold water is required in a waste treatment plant. Though some design elements are common to the MBBR process, the design of an IFAS is more complex given the recirculation of secondary sludge in the reactor.

The hybrid process with a fluidized bed placed in an activated sludge reactor (IFAS) is a technology for the treatment of carbon and nitrogen that is very attractive due to its compactness and easily adaptable for the rehabilitation of activated sludge water treatment plants.

The IFAS process combines the conventional technologies of activated sludge and biofilms in a single reactor. Although an IFAS system varies in terms of components, a typical IFAS configuration is similar to that of an activated sludge plant, involving the recirculation of activated sludge at the head of the biological tank. This configuration leads to the operation with two different biological populations acting in synergy. Mixed liquor suspended solids (MLSSs) degrade most of the organic load, and the biofilm establishes a strong nitrifying population, allowing the oxidation of the nitrogen load. Due to the biofilm processes, anaerobic, aerobic and anoxic zones can exist together in a single stage. The IFAS process allows a biological population to be fixed on a support, thus meaning it is no longer necessary to increase the population of free cultures.

The treatment system including the IFAS process involves a treatment-appropriate design. Fine screening and primary settling can be used to prevent the accumulation of materials and the abrasion of supports dispersed in the aeration tank as well as their clogging.

IFAS technology is equally recommended for new constructions and for upgrading already existing installations, both in municipal and industrial installations. Fixed cultures can be used to reduce the dimensions of tanks that are designed to receive populations that may be different and distributed in free culture in the tank and fixed on a mobile support.

When they are chosen to improve existing installations, IFAS systems increase the treatment capacity without requiring an increase in the tank volume. Before installing an IFAS system in an existing plant, the aeration capacity is evaluated to make sure it is sufficient to manage the increased biochemical removal of BOD and the respiration of the biomass resulting from the improvement of treatment performances provided by IFAS. The installation of supports also requires serious inspection to ensure that the tanks are able to properly accommodate equipment and biological supports. The global biomass in the IFAS increases sludge age, facilitating better nitrification compared to simple suspended growth systems. Nitrification is also more rapidly restored as a large amount of nitrifying agents is preserved on biological supports. IFAS technology allows the addition of biomass in a wastewater treatment installation in order to meet stricter parameters or increased loads without the need to have available additional tanks.

It is known that the improvement of wastewater treatment plants usually consists of increasing the volume of the bioreactor in order to develop an additional bacterial population required to address the load needs. However, the limits of the load of solids acceptable for the existing clarifier prevent the addition of an amount of sludge that would exceed the acceptable invert load that could be transported in the suspended growth system. With an IFAS, the additional biomass provided by the biofilm does not increase the MLSS load of the clarifier, and it is a factor that often limits the treatment capacity of conventional activated sludge, systems.

Compared to activated sludge processes, biofilm processes have the advantage that anaerobic, anoxic and aerobic zones can be combined in a single stage. IFAS systems allow the additional bacterial population to exist on a fixed surface, thus eliminating the need to increase the suspended growth population.

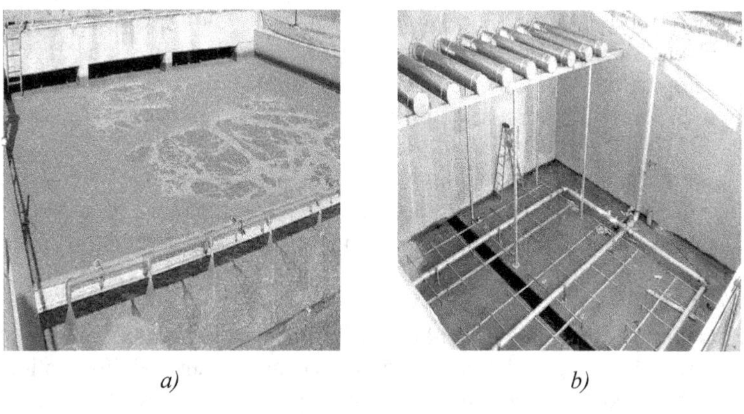

a) *b)*

Figure 2.22. *IFAS: a) operating and b) empty*

2.3.2. *Materials*

There are a certain number of different approaches for the implementation of IFAS systems, but the various configurations can be grouped into two fundamental types: dispersed media, trapped in the aeration tank, and fixed media, such as structured sheet supports or knitted fabric supports, fixed in place in the aeration tank.

In general, an IFAS configuration will resemble that of an activated sludge plant, with biomass carriers introduced in carefully selected zones within activated sludge processes. This allows two distinct biological populations to act synergistically, the MLSS degrading the major part of the organic load (BOD), and the biofilm creating a highly nitrifying population for the oxidation of the nitrogen load (NH_4^+).

The interest gained by the IFAS process is due to the possibility of removing organic carbon and nitrogen in the same reactor, the reduction of the footprint, the resistance to hydraulic shocks and the short reaction times in the biological tank.

IFAS technology meets the need to increase the capacity of activated sludge tanks with little or no additional tank volume due to the fixed biomass, which provides purifying microorganisms without modifying the global volume.

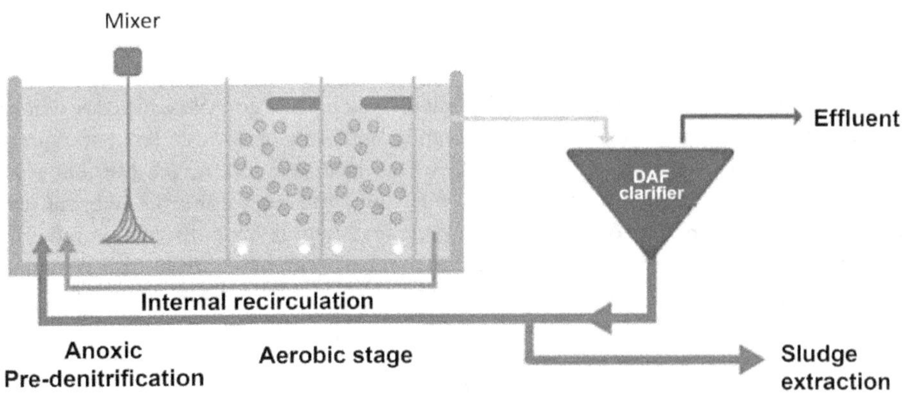

Figure 2.23. *IFAS process diagram: activated sludge*

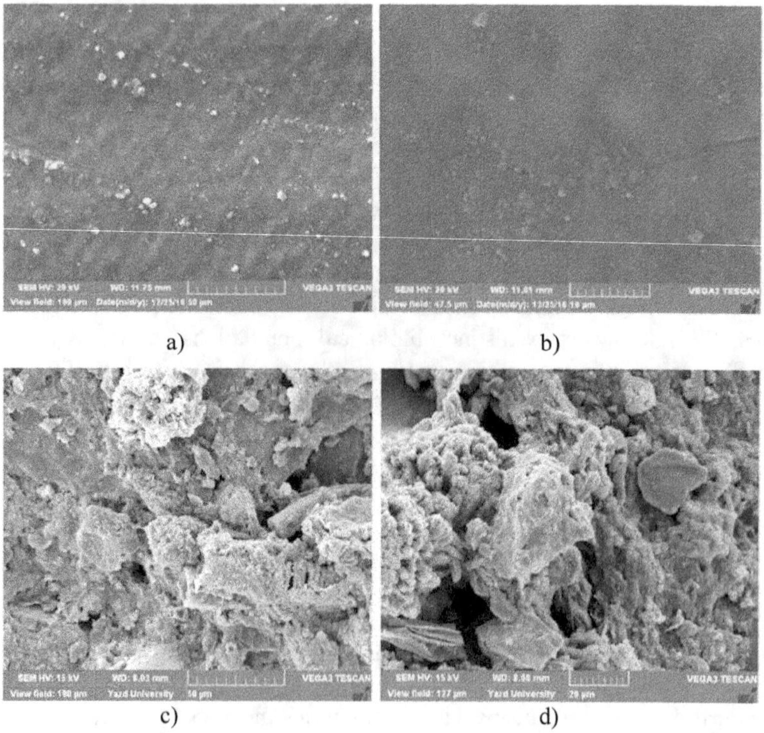

Figure 2.24. *Electron microscope images of biological tanks: a,b) without support; c,d) with supports in an IFAS (Eslami et al. 2018)*

However, the IFAS process as a biological system is a complex process whose effectiveness is influenced by various operating and structural variables, particularly MLSSs, the HRT, the type and filling rate of the biofilm support, the nutrient with respect to the microorganism, the air flow rate, the temperature, the pH, the nutrient levels (nitrogen and phosphorus) and the organic load rate. IFAS optimization requires a lot of time and efforts to obtain higher removal of pollutants while also reducing energy consumption.

2.3.3. *Implementation and operation*

Similarly, in renovation applications, an increased treatment capacity can be reached, with the other advantages of fixed film processes. The existing aeration capacity must be evaluated to determine whether it is adequate for the increase in BOD removal and in the expected biomass respiration with the highest performance level of the IFAS-associated treatment.

The IFAS process is considered one of the cutting-edge processes, offering a sustainable solution for wastewater treatment. IFAS is presented as a technical advancement in moving-bed biofilm processes by integrating the systems with free and fixed cultures. As a general rule, IFAS reaches a removal rate above 90% for COD and ammonia, improves sludge settling properties and strengthens operational stability.

The biological treatment of wastewater is a process that uses microorganisms that transform a large part of biodegradable organic materials into CO_2, H_2O and biomass.

IFAS can achieve improved biological removal of nitrogen thanks to nitrifying agents fixed on the support that also facilitate the compactness and robustness of the system. Coexisting biofilm and flocs accommodate simultaneously autotrophic and heterotrophic bacteria, which improves the removal of phosphorus and nitrogen. IFAS achieves nitrification throughout the year and high treatment effectiveness thanks to the robustness of the biofilm, allowing IFAS to operate with very long sludge retention times (SRTs). IFAS maximizes the use of the volume of the bioreactor, unlike other attached growth processes such as rotating biological contactors (RBCs) that only use part of the volume of the bioreactor (Waqas et al. 2020). Hence, the construction characteristics of the attached and suspended growth systems are incorporated in the IFAS system (Leyva-Diaz et al. 2017).

A limited body of literature is currently available on IFAS. Ødegaard (2006) examined MBBRs for the removal, nitrification and denitrification of organic materials. The biofilm and the flocs, both present in IFAS, are two common microbial communities that feature distinct properties. Their structural and physicochemical properties indicate their role in the degradation of pollutants (Mahendran 2012). For example, in the case of activated sludge, the floc is compact (denser than water) and is easily deposited to produce an effluent nearly without sludge. Conversely, the sludge detached from the biofilm is not easily settled because of its physico-chemical properties.

2.3.4. *IFAS operating parameters*

Biofilm growth is influenced by a certain number of operating and environmental conditions, particularly the biodegradable soluble COD/BOD_5 load, SS concentration, temperature and concentration of DO. The fixed biomass reacts rapidly when the load conditions are modified in the system. The performances of pilot and large-scale IFAS systems have been widely reported. It is important to note the role of parameters such as the carrier medium, the HRT, the aeration rate, the SRT, the DO and the filling rate of the reactor. The biological performances of IFAS in terms of biological nitrogen removal (BNR) and phosphorus removal are very encouraging (Di Trapani et al. 2011, 2013; Massé and Comeau 2021).

The three most important parameters of the IFAS concern the removal of organic materials.

COD removal varies between 64.8 and 90%. Stable nitrification is generally obtained by this process, corresponding to about 50% of the volume required in a conventional system and at higher loading rates. The biomass of the fixed film represents up to 50% of the total biomass of the system. Sludge age ranges between 5 and 7 days. The free culture biomass has an aerobic sludge age of 3.6 days, and the fixed biomass ensures a large part of the nitrification activity (Eslami et al. 2018) at lower temperatures (< 12°C).

2.3.4.1. *Aeration*

The aeration demands of IFAS systems must be considered both from the perspective of investment costs and long-term operational costs. Suspended phase oxygen concentrations must be above the normal (which is 3 to 4 $mg.L^{-1}$) to eliminate any kinetic limitation associated with oxygen diffusion through the biofilm layers. The required quantity of air is also influenced by the effectiveness of the oxygen transfer from the aeration device used for the installation. During a pilot project, higher air flow rates compared to conventional processes were observed. There was no proof of oxygen transfer effectiveness higher than expected compared to the design values used for the test.

2.3.4.2. *Foam*

Given that floating fluids must be retained in activated sludge tanks, natural foam traps are often created in IFAS systems. As a general rule, the only way to fight against foam is to use antifoam sprays and/or to provoke a temporary waste of the surface foam of the system. During the pilot project, the operational and maintenance personnel of the plant developed a system of screens with vertical bars installed near the water surface that managed to retain the media during the foam passage. The screen had openings of ¼ of inch and was supported at the top and at the bottom. A spray nozzle was directed toward the bars, and the screen design was modified to allow tilting in the flow direction.

A new version of IFAS has been presented, H-IFAS, in which the aerobic, anaerobic and anoxic zones are designed and built in a single reactor to reach a complete relative capacity for nitrification, denitrification and phosphorus removal and also a high effectiveness in removing carbon organic compounds. Ozonation is simultaneously used in the aerobic zone to increase the effectiveness of Kjeldahl nitrogen, COD and BOD_5 removal. Meanwhile, due to the combination of various zones of the reactor and its vertical configuration, the dimensions of the global reactor are still acceptable (Andreottola 2003).

Related to the results obtained, the effectiveness of carbon organic material removal from the H-IFAS reactor is much higher (at least five to six times) than the effectiveness of the conventional activated sludge process. Applying biofilm supports in the reactor tends to increase the population of autotrophic microorganisms that play an important role in nitrification, denitrification and biological removal of phosphorus. Phosphorus removal in this reactor is ensured by returning the settled sludge from the clarifier to the lower part of the reactor, which is an AMBR. As proven, the P removal process in this reactor has good stability despite the increase in hydraulic loading rates (HLRs) over time. However, the effectiveness of the reactor compared to other conventional processes is very different. Namely, though the volume of the reactor is limited and the compactness of the system leads to very low dimensions, the capacity of the process in this reactor is reliable, and it has very good effectiveness for eliminating phosphorus from raw wastewaters.

There are several major factors that influence nitrification kinetics. These are the applied organic loads, the hydraulic loads, the temperature, the pH, the concentration of DO and the type of support material.

2.3.5. *IFAS advantages and disadvantages*

Some of the advantages of using IFAS include:

– improvement of nitrification compared to the free culture processes due to the combination of aerobic, anaerobic and anoxic zones and the increase in the SRT;

– more rapid restoration of system nitrification due to the large mass of nitrifiers on the fixed biofilm;

– higher stability of performances;

– small variation in the sludge index (SI);

– reduction in sludge production;

– improvement of denitrification process;

– compatibility with existing fine bubble aeration systems, which contributes to reducing long-term operating costs;

– resistance to organic and hydraulic shocks;

– increased capacity of activated sludge systems in the same tank volume.

Despite the significant advantages, IFAS systems are also accompanied by several physical demands that should be carefully considered during the design and operation of these systems:

– IFAS systems have complex and highly mechanized configurations. The proper operation of these systems requires expertise and well-trained operators.

– An adequate mix must be provided to make sure that floating media remain uniformly distributed.

– The mixing energy provided by the aeration system is essential to biomass desquamation and to the creation of a thin biofilm.

– Aeration for nitrification systems: a higher concentration of DO (generally 3 to 4 mg/L) is required in the suspension phase to ensure the biofilm is completely aerobic.

– To retain the biological supports, effluent screens must be installed to contain the media inside the reactors.

– Foam accumulation is a common problem for IFAS systems.

– Higher operating cost.

2.3.6. *MBBR versus IFAS*

The design and engineering of MBBR and IFAS wastewater treatment plants are fraught with difficulties and the errors made may be difficult to correct. Problems may occur when inexperienced installations, manufacturing companies and plastic manufacturers with no knowledge of process technology try to copy this technology.

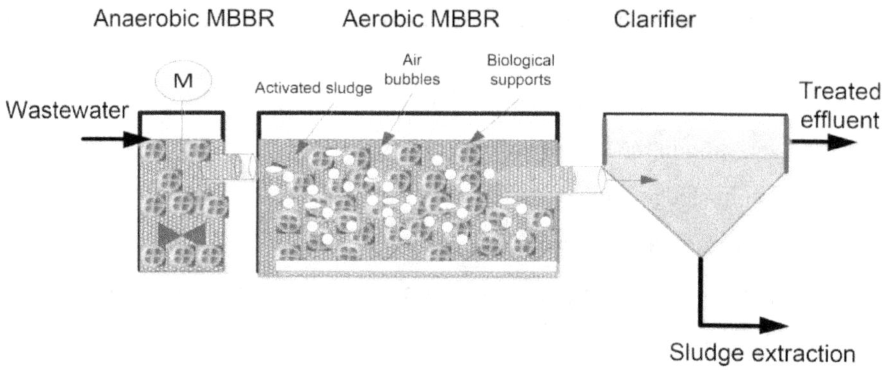

Figure 2.25. *MBBR process*

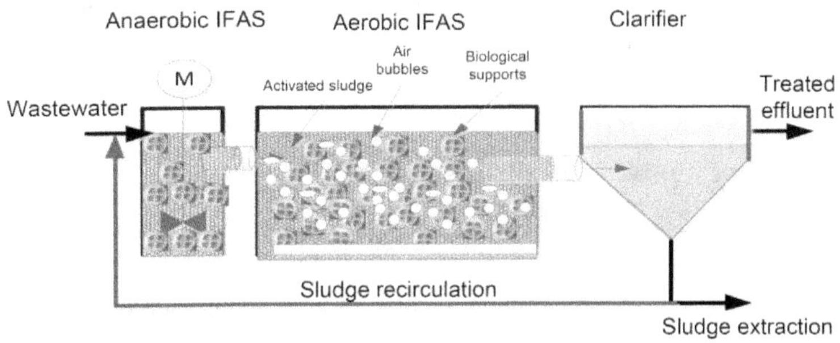

Figure 2.26. *IFAS process*

MBBR is the abbreviation of "Mobile Bed Biofilm Reactor", while IFAS stands for "Integrated Fixed-Film Activated Sludge". This may apparently look like a simple concept: a biological tank is filled with a plastic support that is mixed with a mass of microorganisms and supplied with wastewater. The difference between IFAS and MBBR is that activated sludge resulting from sludge recirculation is used in the IFAS system, meaning that the combination of activated sludge and biofilms fixed on supports is used in the same reaction volume.

The size of MBBR/IFAS tanks and the filling rate are determined as a function of polluting loads and taking into account product factors specific to supports. The crucial factors for the dimensioning of MBBR/IFAS installations are temperature, efficiencies and the specific loading rates related to the selected support.

For example, for the low temperature factor, it is the lowest value that is taken into account for calculating the required volume of tanks and that of supports. The highest temperature is taken into account for the calculation of O_2 requirements.

The specific surface of supports is not sufficient to estimate the biological efficiency, but the total mass of the fixed biofilm is an additional parameter that differs between the two processes. All the more so as the filling rate differs between the two systems, 67% for MBBR and < 50% for IFAS.

For new installations, IFAS generally requires a smaller volume tank than that of conventional activated sludge and would therefore have a lower capital manufacturing cost than a conventional system.

For the modernization of existing activated sludge plants and in order to increase their capacity or improve nitrogen removal, IFAS represents an economic choice associated with the additional volume that would be necessary to obtain the additional capacity of the activated sludge.

IFAS processes require low or no additional operating costs or operators compared to classical treatment systems. Nevertheless, oxygen requirements may be more significant.

2.4. Hybrid plant system: Organica FBR™

2.4.1. *Operating principle*

Organica FBR™ can be considered similar to a sequencing batch reactor (SBR) process with a separate anoxic zone, allowing it to beneficially combine the SBR operating simplicity with the denitrification capacity of a classical anoxic zone. This technology also relies on the biomass fixed on the plant root system to increase the purifying efficiency. Roots are not likely to clog; therefore, the operation is safer.

Figure 2.27. *Operating diagram of the Organica™ process*

The use of a set of plants as bacteria support is one of the major specificities of technology. Roots are immersed in the effluent at a depth between 0.5 and 1 meter and are a healthy habitat for living organisms ranging from bacteria to snails. Plants excrete a small quantity of organic acid on the root surface feeding the biofilm. Due to this symbiosis the biomass is maintained active even when the liquid level in the tank is low. This is how plants help bacteria stay alive during periods of scarcity, and there will be growth of bacteria in the system when the liquid level is maximal. Finally, Organica™ technology is expected to provide high flexibility and better treatment stability than a conventional system.

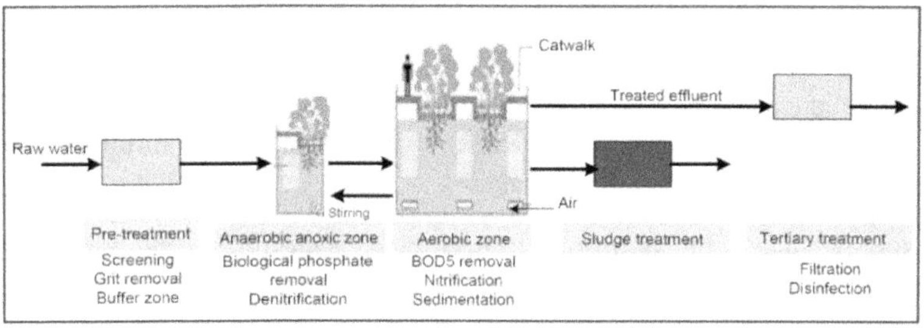

Figure 2.28. *Diagram of a biological treatment system with the Organica process*

The plant environment is a set of textile fibers allowing protection of the roots against aeration and also the development of an additional surface for the biofilm.

As water treatment is cyclic, taking place in an SBR, each stage can be identified by specific operating parameters:

– supply stage;

– reaction stage;

– settling stage;

– sedimentation/drawing stage.

Each treatment stage is associated with a series of settings that improve the treatment quality while optimizing energy consumption. The system setups are installed in a greenhouse offering the best possible conditions for plant growth. This structure also allows the operator to be protected from weather changes. The treatment row also features a deoxygenating vent, allowing the reduction of the DO rate before the effluent passes to the anoxic tank. The deoxygenating process should allow the consumption of the 2 g of oxygen/m³ present in the aeration tank. The foams formed at the surface of reactors are collected in a tank installed near the deoxygenating compartments.

A global treatment system includes:

– a pretreatment;

– an anaerobic tank and/or an anoxic tank;

– an aeration tank operating in sequences with a sedimentation;

– a sludge treatment;

– a tertiary treatment, if needed.

Pretreatment may consist of a rotary screen that screens effluents at 1,000 μm, allowing excellent screening and recovery of sands and greases, at least equivalent to that of classical pretreatments. This waste is then transported through a perpetual screw to a compacting, dewatering zone.

The supply stage is the start point of each cycle. During this stage, raw effluent is transferred from the buffer tank to the anoxic tank. Then, the effluent circulates in the aeration tank (via the recirculation pump) and in the deoxygenating vent. There may be sequential aeration during this stage.

The start and the end of the filling stage are controlled by the water level in the aeration tank before and after filling. Supply starts when the buffer tank contains enough effluents to completely fill the aeration tank.

The reaction stage is essential, as during this stage pollution is treated. Global control of nitrification, denitrification and degradation of pollutants is performed by the operator or by an automatic control system that sets the parameters. Recirculation pumps installed between the aeration tanks and the anoxic tanks perform the effluent transfer required for the complete treatment of nitrogen (nitrification/denitrification).

The recirculation of nitrate-rich wastewater is directed to the anoxic tank where denitrification takes place. The metabolic mechanism of denitrifying bacteria uses the organic materials (BOD_5) brought by raw waters that supply the anoxic tank.

Fine bubble diffusers associated with air boosters ensure the oxygenation of the aeration tank. The foams produced during this stage are evacuated to the foam collector. Recirculation ensures very good denitrification efficiency. Aeration in the aerated zone can be intermittent, which provides a complementary anoxic volume to improve denitrification. Physico-chemical removal of phosphate may contribute to the phosphorus removal by the injection of ferric chloride in the aerated tank.

The settling stage starts at the end of the biological reaction. At the beginning of this stage, all equipment is shut down. Sludge settling allows the extraction of treated water at the surface. The settling duration ranges between 30 and 60 minutes in general. The sludge extraction stage is the fourth and the last stage of the treatment cycle. To keep a constant mass of sludge in the system, produced sludge is regularly evacuated at each cycle to a thickener. Tertiary treatments can be added to the system: complementary chemical precipitation, filtration on material (sand or active carbon), disinfection of membranes.

Phosphorus pollution is treated by simultaneous precipitation. This involves the introduction of the reagent (ferric chloride $FeCl_3$) directly at the level of the aeration tank. The capacity of the storage tank is 20 m^3. Two membrane volumetric metering pumps are used to inject ferric chloride in the aerated tanks.

Over 50% of the surface of anoxic tanks and aeration tanks is covered by plants cultivated in half-submerged containers (5 plants/m^2). Containers consist of a support grid and a fine mesh to retain the media (clay balls in which plants are planted).

In order to control the development of the plants of the ecosystem, a greenhouse made from horticultural glass covers the entire surface of the biological tanks. The ridge height is 5–5.5 m, and the free passage height is 4 m. The greenhouse maintains high temperature and hygrometry above the plants; it also offers the operator pleasant working conditions, irrespective of the period of the year. Greenhouse temperature must be maintained above 6°C. When the temperature drops, a temperature-controlled backup heating system is used and regulates temperature between 6 and 10°C.

In order to permanently have an industrial water reserve that can be used for cleaning operations or for automated washing of equipment (screen and centrifuge), a storage tank is placed upstream of the flow rate measurement channel, whose volume must allow washing operations for 3 h (critical case in dry time). A sampling point is placed on the flow meter channel installed at the station outlet and is also flow rate-controlled.

The sludge treatment station includes a thickener that allows pre-thickening and homogenization of sludge due to the installation of an agitator. The overflow of the overflow system can be sent to the head of the system. In some cases, the thickener is covered and deodorized using an activated carbon filter. Extracted sludge is sent to dehydration (centrifugation, filter press) with polymer injection.

Dehydrated sludge is sent to a storage zone, drying beds, reed beds or other technologies.

2.4.2. *Performances*

Similar to any other wastewater treatment station, the performance parameters are monitored based on pollution flow rates and concentrations of the inlet flux and outlet flux. Operating parameters are also reported: concentration in the tanks, V30, SI (Mohlmann's index), sludge age, aeration time, aerated sludge age, consumption of reagent for phosphate removal ($FeCl_3$), energy consumption, records of various temperatures in the tanks or under the greenhouse, humidity, motor operating time and specific flow rates.

Finally, punctual analyses (water and sludge) are conducted at various stages of the system: analyses of sludge treatment returns, analyses of the biomass fixed on the roots (activity tests), measuring odors in the greenhouse, measuring emerging pollutants (micro pollutants, pharmaceutical pollutants).

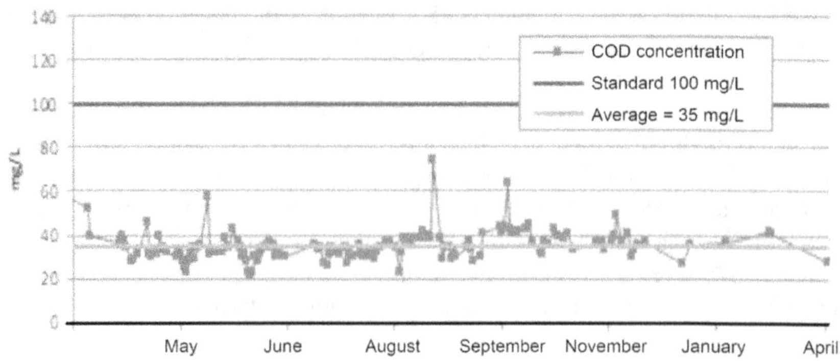

Figure 2.29. *Evolution over one year of outlet COD in an Organica station. For a color version of this figure, seewww.iste.co.uk/gaid/wastewaterengineering2.zip*

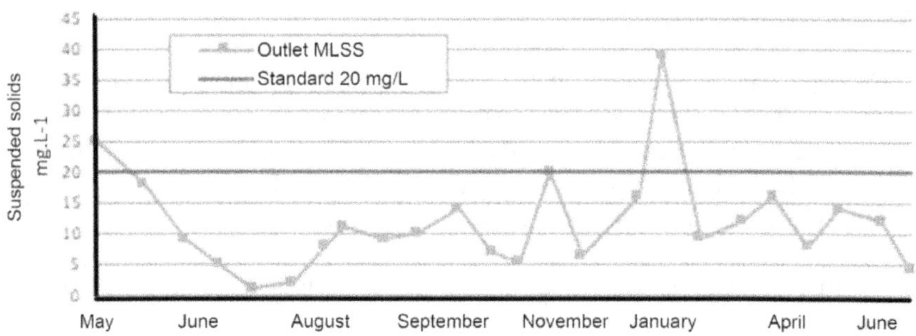

Figure 2.30. *Evolution over one year of outlet suspended solids. For a color version of this figure, seewww.iste.co.uk/gaid/wastewaterengineering2.zip*

According to the results obtained from about 30 stations, BOD_5 is removed at over 95% with outputs below 20 $mg.L^{-1}$ and COD is removed at over 90% with outputs below 50 $mg.L^{-1}$.

Figure 2.29 shows the set of on-site measured values and not only those of compliance assessment.

The concentration of SS in the effluent is most of the time below the effluent standard of 20 $mg.L^{-1}$.

2.4.2.1. *MLSS concentration in the biological tank*

The average concentration of MLSS in the two tanks of the station is on average around 3.5 $g.L^{-1}$. Considering about 75% VSS in the MLSS, the resulting concentration is 2.6 $g.L^{-1}$.

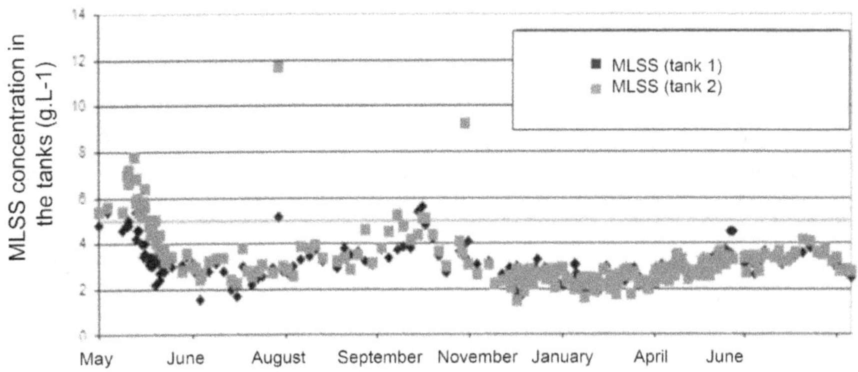

Figure 2.31. *MLSS concentration in biological tanks. For a color version of this figure, seewww.iste.co.uk/gaid/wastewaterengineering2.zip*

2.4.2.2. *Sludge age*

Sludge age represents the biomass retention time in the treatment system. It is calculated with the following formula:

$$\frac{\text{Conc. tank (g/L)} * \text{V tank (m}^3)}{\text{Conc. ext. sludge (g/L)} * \text{V ext. (m}^3\text{/d) x 0.8} + \text{output MES}\left(\frac{\text{mg}}{\text{L}}\right) * \text{extracted volume (m}^3\text{/d)}}$$

The role of the 0.8 coefficient is to take into account the suspended solids that return to the tanks after sludge treatment. The sludge ages in the two aerated tanks follow approximately the same variations. The average sludge age is 18 days, which corresponds to the plant operating under low load. The age of aerated sludge corresponds to the sludge age reported during the duration of stages known as "active", namely supply, reaction and settling. The average duration of these stages over one day is about 20 h. For this, the formula becomes:

$$\text{aerated sludge age} = \text{total sludge age (d)} * \frac{\text{volume of aeration tank (409.2m}^3)}{\text{volume of total tank (654.5m}^3)} * \frac{\text{aeration duration of the tanks (h)}}{20\ \text{h}}$$

On average, aerated sludge age is 5 days; fluctuations can be noted between 2.2 days and 10 days. Although the aeration duration is not constant, it can be noted that the variations of aerated sludge age follow exactly those of sludge age.

2.4.2.3. *V30 and SI*

V30 (volume settled after 30 minutes) is a first indicator of sludge settleability. The lower the V30, the better the settling. This allows checking that there are no floating agents (filamentous bacteria), or that the water in the tanks does not remain cloudy after 30 minutes of settling.

The SI (or Mohlmann's index, MI) is the main indicator of sludge settleability. It serves as a reference:

– MI < 50 mL.g^{-1}: poor settleability;

– MI $\in$ 80–150 mL.g^{-1}: good settleability;

– MI > 150 mL.g^{-1}: bulking phenomenon (filamentous bacteria).

Most of the time, SI ranges between 80 and 150 mL.g^{-1}.

2.4.2.4. *Sludge production*

Organica announces a 20% reduction of sludge production compared to a standard activated sludge treatment station.

Sludge is monitored in several points of the station: in the tanks, upon sludge extraction to the thickener and upon coming out of this thickener to the centrifuge.

Table 2.7 presents the results obtained.

	VSS concentration in g.L^{-1}	
	Row 1	**Row 2**
Tanks	2.75	2.71
Extraction	9.52	9.96
Centrifuge feed	17.4	

Table 2.7. *VSS concentration in various compartments*

Sludge production is given by the following relations:

Produced sludge = stored sludge + 0.8 × extracted sludge

Stored sludge = tank concentration × tank volume

Extracted sludge = extracted sludge concentration × extracted volume

The sludge production coefficient is 0.43 kg MES. kg^{-1} treated COD.

2.4.3. *Invasive risks of plants*

Two main types of biomass can be identified: free biomass in the tanks and biomass fixed on roots. According to bacteriological analyses, free bacteria in aerated and anoxic zones are similar, in terms of quantity of microorganisms and bacterial species composition. This can be explained by the recirculation of sludge from aerated tanks to anoxic tanks.

The Organica FBR™ process is a treatment system combining ecological engineering with traditional wastewater treatment technologies. The technology relies indeed on activated sludge treatments, associating aeration and settling tanks with plants that serve as support to a fixed culture (on their roots). A list of plants that can be used in Organica processes is provided by the Organica company.

Most of these plants are of tropical origin, used for their ornamental properties and their aptitudes to develop roots as support for the biomass; the question that arises is that of their associated invasive risk.

Indeed, some exotic species (introduced by humans either voluntarily or not, and which are beyond their distribution area) can proliferate and cause damage to biodiversity, human health, etc. These are then referred to as invasive species.

An expert opinion was demanded from the Muséum national d'Histoire naturelle (MNHN) (French National Museum of Natural History) and the Fédération des Conservatoires botaniques nationaux (FCBN) (French Botanical Conservatories).

According to the consultation conclusions, certain species can be banned in some countries, in line with local regulations. For example, species such as *Amorpha fruticosa*, *Arundo donax*, *Fraxinus pennsylvanica*, *Miscenthus sinensis* and *Zantedeschia aerthiopica* are not recommended in France. The 2,000 selected and available plants offer a wide range of choices. A charming reconstruction of a lush garden may help local residents ignore the prosaic reality of a wastewater treatment plant.

Amorpha fruticosa

Arundo donax

Fraxinus pennsylvanica

Miscenthus sinensis

Figure 2.32. *Plants not recommended in France*

The Organica wastewater treatment technology is based on a plant ecosystem composed of attached biofilm developing on the roots of tropical plants, 20 cm under water in a compact wastewater treatment plant that can be covered by a greenhouse.

The wastewater treatment plant operates according to the biological process of the SBR or of the fluidized bed reactor (FBR). This process is currently one of the most compact available, after the wastewater treatment by membrane processes. The FBR process relies on the alternating zone principle.

Various types of microorganisms are fixed on the surface of roots, thus reducing the pollutant load and decontaminating water. The specific surface of the plant root system is particularly important, as it is much larger than the surface of the artificial substrate, which leads to a higher concentration of the available fixed biomass.

Figure 2.33. *Greenhouse protecting plants and ensuring a homogeneous temperature*

Figure 2.34. *Greenhouse ceiling at a height between 5 and 5.5 m*

Figure 2.35. *Aquatic plants contributing to wastewater treatment*

At the same time, biofilm activity feeds on the oxygen present in the interstitial fluid. The plants secrete a small quantity of organic acid at the surface of roots that serves as a nutrient for the biofilm. The symbiosis between the root and the complex eco-system allows the biomass to remain active even when there is a low level of liquid in the tank. The plants also help bacteria to live during scarcity periods (loading variation in the wastewater treatment plant). The solution is the implementation in a temperate climate, adapted to an urban environment: this solution is compact (half a classical STEP) and aesthetic; it is odorless thanks to the plant cover; the robustness of wastewater treatment is strengthened by the buffering power of vegetation; and due to the above characteristics, plants can be grown.

Wastewater treatment plants with lush vegetation look like tropical greenhouses. Being space and energy efficient, competitive on the mid-range market and instructive for the general public, they are attracting the interest of major water players due to the new perspectives they offer on the sanitation market. This process can be used in educational programs.

2.5. References

Andreottola, G., Foladori, P., Gatti, G., Nardelli, P., Pettena, M., Ragazzi, M. (2003). Upgrading of a small overloaded activated sludge plant using a MBBR system. *J. Environ. Sci. Health A Tox. Hazard Subst. Environ. Eng.*, 38(10), 2317–2328.

Brosseau, C. (2015). Récupération de la matière organique biodégradable présente dans l'effluent d'un MBBR à forte charge. Thesis, Université de Montréal, Montreal.

Canler, J.P. (2011). Procédés MBBR (Moving Bed Biofilm Reactor). Cas du procédé R3F. Technical Data Sheet, Cemagref, Paris.

Chattopadhyaya, I., Rajesh Banu, J., Mohamed Usmanb, T.M., Varjani, S. (2022). Exploring the role of microbial biofilm for industrial effluents treatment. *Bioengineered*, 13(3), 6420–6440.

Daigger, G.T. and Boltz, J.P. (2018). Oxygen transfer in moving bed biofilm reactor and integrated fixed film activated sludge processes. *Water Environ. Res.*, 90(7), 615–622.

Delatolla, R., Tufenkji, N., Comeau, Y., Gadbois, A., Lamarre, D., Berk, D. (2010). Investigation of laboratory-scale and pilot-scale attached growth ammonia removal kinetics at cold temperatures and low influent carbon. *Water Quality Res. J.*, 45(4), 427–436.

Di Trapani, D., Christensson, M., Ødegaard, H. (2011). Hybrid activated sludge/biofilm process for the treatment of municipal wastewater in a cold climate region: A case study. *Water Sci. Technol.*, 63(6), 1121–1129.

Di Trapani, D., Christensson, M.D., Torregrossa, M., Viviani, G., Ødegaard, H. (2013). Performance of a hybrid activated sludge/biofilm process for wastewater treatment in a cold climate region: Influence of operating conditions. *Biochem. Eng. J.*, 77, 214–219.

Eslami, H., Ehrampoush, M.H., Falahzadeh, H., Hematabi, P.T., Khosravi, R., Dalvand, A., Esmaeili, A., Taghavi, M., Ebrahimi, A.A. (2018). Biodegradation and nutrients removal from greywater by an integrated fixed-film activated sludge (IFAS) in different organic loadings rates. *AMB Express*, 8(3), 25–39.

Federico, J., Richtarik, S., Gilligan, T. (2002). IFAS system demonstration of cold temperature nitrogen removal. In *Proceedings of the 75th Annual Water Environment Federation Technical Exhibition and Conference*. Water Environment Federation, Alexandria.

FNDAE (2007). Présentation générale du procédé R3F. Report, FNDAE, 34.

Germain, E., Bancroft, L., Dawson, A., Hinrichs, C., Fricker, L., Pearce, P. (2007). Evaluation of hybrid processes for nitrification by comparing MBBR/AS and IFAS Configurations. *Water Sci. Technol.*, 55(8–9), 43–9.

Helness, H., Melin, E., Ulgenes, Y., Jarvinen, P., Rasmussen, V., Odegaard, H. (2005). High-rate wastewater treatment combining a moving bed biofilm reactor an enhanced particle separa-tion. *Water Sci. Technol.*, 52(10–11), 117–127.

Hoang, V., Delatolla, R., Abujamel, T., Mottawea, W., Gadbois, A., Laflamme, E., Stintzi, A. (2014). Nitrifying moving bed biofilm reactor (MBBR) biofilm and biomass response to long-term exposure to 1 °C. *Water Res.*, 49, 215–224.

Horn, H. and Hempel, D.C. (1995). Mass transfer coefficients for an autotrophic and a heterotrophic biofilm system. *Water Sci. Technol.*, 32(8), 199–204.

Hubbell, S.B., Pehrson, R.L., Schuler, A. (2006). Eight years of successful cold weather nitrification with integrated fixed-film/activated sludge. In *Proceedings of the 79th Annual Water Environment Federation Technical Exhibition and Conference*. Water Environment Federation, Alexandria.

Johnson, C.H. and Haegh, M. (2005). Increasing BNR capacity in activated sludge systems by use of suspended carrier IFAS technology – Example from a full scale installation at Broomfield BWRF in Colorado, USA. In *Proceedings of the IWA Specialist Conference Nutrient Management in Wastewater Treatment Processes and Recycle Streams*. IWA, London.

Johnson, T.L. and McQuarrie, J.P. (2002). IFAS BNR full-scale design and performance challenges. In *Proceedings of the 75th Annual Water Environment Federation Technical Exhibition and Conference*. Water Environment Federation, Alexandria.

Jones, R.M., Lambert, R., Manoharan, R., Crane, R., Campbell, H. (1999). Demonstration of an integrated fixed film activated sludge (IFAS) system for enhanced nitrification using a new free moving media. In *Proceedings of the Annual Technical Symposium of the Water Environment Association of Ontario*. Water Environment Association of Ontario, Milton.

Kaldnes Miljøteknologi, A.S. (2011). Design manual. AnoxKaldnes.

Leyva-Diaz, J.C., Martin-Pascual, J., Povatos, J.M. (2017). Moving bed reactor to treat wastewater. *Inter. J. Environ. Science & Technology*, 14, 871–910.

Maas, C.L.A., Parker, W.J., Legge, R.L. (2006). Process observations from start-up of an integrated fixed-film activated sludge process. In *Proceedings of the 35th Annual WEAO Technical Symposium and OPCEA Exhibition*, Toronto.

Mahendran, B., Lishman, L., Liss, S.N. (2012). Structural physicochemical and properties of flocs ans biofilms in integrated fixed-film activated sludge (IFAS) systems. *Water Res.*, 46(6), 5085–5101.

Massé, M. and Comeau, Y. (2021). Guide pour l'étude des technologies conventionnelles de traitement des eaux usées d'origine domestique, réacteur biologique à garnissage en suspension (RBGS). Guide, Government of Québec, Quebec.

McQuarrie, J.P. and Boltz, J.P. (2011). Moving bed biofilm reactor technology: Process applications, design, and performance. *Water Environ. Res.*, 83(6), 560–575.

McQuarrie, J.P., Rutt, K., Seda, J., Haegh, M. (2004). Observations from the first year of full-scale operation – The IFAS/BNR process at the Broomfield wastewater reclamation facility, Broomfield, CO. In *Proceedings of the 77th Annual Water Environment Federation Technical Exhibition and Conference*.

Melin, E., Helness, H., Kenakkata, T., Ødegaard. H. (2004). High-rate wastewater treatment based on moving bed biofilm reactor, polymer coagulation and flotation. In *Chemical Water and Wastewater Treatment VIII*, Hahn, H.H., Hoffmann, E., Ødegaard, H. (eds). IWA Publishing, London.

Metcalf & Eddy Inc. (2014). *Wastewater Engineering: Treatment and Resource Recovery*, 5th edition. McGraw-Hill, New York.

Moretti, P. (2015). Performances, modélisation et limites d'un procédé à lit fluidisé associant culture libre et fixée (IFAS) pour le traitement du carbone et de l'azote des eaux résiduaires. PhD Thesis, Claude Bernard University Lyon 1, Lyon.

Moretti, P., Canler, J.-P., Petrimaux, O. (2016). Performances et limites d'un procédé à lit fluidisé associant biofilm et liqueur mixte (IFAS) pour le traitement du carbone et de l'azote dans les eaux usées résiduaires. *Eaux urbaines*, 11. doi: 10.1051/tsm/201611016.

Morgenroth, E. (2008). Modelling biofilms. In *Biological Wastewater Treatment – Principles, Modeling, Design*, Henze, M., van Loosdrecht, M.C.M., Ekama, G., Brdjanovic, D. (eds). IWA Publishing, London.

Morgenroth, E. and Wilderer, P. (2000). Influence of detachment mechanisms on competition in biofilms. *Water Res.*, 34(2), 417–426.

Ødegaard, H. (1999). The moving bed biofilm reactor. In *Water Environmental Engineering and Reuse of Water*, Igarashi, T., Watanabe, Y., Asano, T., Tambo, N. (eds). Hokkaido Press, Sapporo.

Ødegaard, H. (2000). Advanced compact wastewater treatment based on coagulation and moving bed biofilm processes. *Water Sci. Technol.*, 42(12), 33–48.

Ødegaard, H. (2006). Innovations in wastewater treatment: The moving bed biofilm process. *Water Sci. Technol.*, 53(9), 17–33.

Ødegaard, H. (2016). A roadmap for energy-neutral wastewater treatment plants of the future based on compact technologies (including MBBR). *Front. Environ. Sci. Eng.*, 10(4), 1–17.

Ødegaard, H. (2018). MBBR and IFAS systems. In *Advances in Wastewater Treatment*, Mannina, G., Ekama, G., Ødegaard, H., Olsson, G. (eds). IWA Publishing, London.

Ødegaard, H., Rusten, B., Westrum. T. (1994). A new moving-bed biofilm reactor: Applications and results. *Water Sci. Technol.*, 29(10–11), 157–165.

Ødegaard, H., Gisvold, B., Helness, H., Sjøvold, F., Zuliang, L. (2000). High rate biological/chemical treatment based on the moving bed biofilm process combined with coagulation. In *Chemical Water and Wastewater Treatment VI*, Hahn, H.H., Hoffmann, E., Ødegaard, H. (eds). Springer, Berlin, Heidelberg.

Onnis-Hayden, A., Dair, D., Johnson, C.H., Schramm, A., Gu, A.Z. (2007). Kinetics and nitrifying populations in nitrogen removal processes at a full-scale integrated fixed-film activated sludge (IFAS) plant. In *Proceedings of the 80th Annual Water Environment Federation Technical Exhibition and Conference*. Water Environment Federation, Alexandria.

Phillips, H.M., Maxwell, M., Johnson, T., Barnard, J.L., Rutt, K., Seda, J., Corning, B., Grebenc, J.M., Love, N., Ellis. S. (2008). Optimizing IFAS and MBBR designs using full-scale data. In *Proceedings of the WEF 81st ACE*.

Piculell, M. (2016). New dimensions of moving bed biofilm carriers: Influence of biofilm thickness and control possibilities. PhD Thesis, Lund University, Lund.

Psaltakis, E.P., Liubicich, J., Pitt, P., Antonio, P., Posada, S., Zabinski, A.J., Lauro, T.J. (2003). Demonstration of integrated fixed film activated sludge process for BNR at the Mamaroneck WWTP. In *Proceedings of the 76th Annual Water Environment Federation Technical Exposition and Conference*. Water Environment Federation, Alexandria.

Rahimi Y., Torabian, A., Mehrdadi, N., Habibi-Rezaie, M., Pezeshk, H., Nabi-Bidhendi, G.-R. (2011). Optimizing aeration rates for minimizing membrane fouling and its effect on sludge characteristics in a moving bed membrane bioreactor. *Journal of Hazardous Materials*, 186(2–3), 1097–1102.

Ross, D. and Fernandes, W. (2002). Innovative integrated fixed-film/activated sludge retrofit. *Environ. Sci. Eng. Mag.* [Online]. Available at: https ://esemag.com/archives/innovative-integrated-fixed-film-activated-sludge-retrofit-could-save-millions/.

Ross, D., Fernandes, W., Briggs, T., Kirn, N., Booth, G., Neely, D., Welp, J. (2004). Integrated fixed film activated sludge (IFAS) at the Lakeview WWTP. The real implementation issues. In *Proceedings of the 77th Annual Water Environment Federation Technical Exhibition and Conference*. Water Environment Federation, Alexandria.

Rusten, B., Hem, L.J., Ødegaard, H. (1995). Nitrification of municipal wastewater in moving-bed biofilm reactors. *Water Environment Research*, 67(1), 75–86.

Rusten, B., Eikebrokk, B., Ulgenes, Y., Lygren. E. (2006). Design and operations of the Kaldnes moving bed biofilm reactors. *Aquacult. Eng.*, 34(3), 322–331.

Rutt, K., Seda, J., Johnson, C.H. (2006a). A successful conversion. *Water Environ. Technol.*, 17(3), 62–66.

Rutt, K., Seda, J., Johnson, C.H. (2006b). Two year case study of integrated fixed film activated sludge (IFAS) at Broomfield, CO WWTP. In *Proceedings of the 79th Annual Water Environment Federation Technical Exhibition and Conference*. Water Environment Federation, Alexandria.

Sander, S., Behnisch, J., Wagner, M., (2017). Energy, cost and design aspects of coarse- and fine-bubble aeration systems in the MBBR IFAS process. *Water Sci. Technol.*, 75(3–4), 890–897.

Sen, D., Copithorn, R.R., Randall, C.W. (2005). Municipal IFAS and MBBR systems as measured in terms of minimum hydraulic retention times and mixed liquor MCRT. In *Proceedings of the 78th Annual Water Environment Federation Technical Exhibition and Conference*. Water Environment Federation, Alexandria.

Soleimani Karizmeh, M. (2012). Investigation of biologically produced solids in moving bed bioreactor (MBBR) treatment systems. Master's Thesis, University of Ottawa, Ottawa.

Sriwiriyarat, T., Pittayakool, K., Fongsatitkul, P., Chinwetkitvanich, S. (2008). Stability and capacity enhancements of activated sludge process by IFAS technology. *J. Environ. Sci. and Health Part a-Toxic/Hazardous Substances Environ. Eng.*, 43(11), 1318–1324.

Stenstrom, M.K., Rosso, D., Melcer, H., Appleton, R., Occiano, V., Langworthy, A., Wong. P. (2008). Oxygen transfer in a full-depth biological aerated filter. *Water Environ. Res.*, 80(7), 663–671.

Stricker, A.E., Barrie, A., Maas, C.L.A., Fernandes, W., Lishman, L. (2009). Comparison of performance and operation of side-by-side integrated fixed-film and conventional activated sludge processes at demonstration scale. *Water Environ. Res.*, 81(3), 219–232.

Torresi, E., Fowler, S.J., Polesel, F., Bester, K., Andersen, H.R., Smets, B.F., Plosz, B.G., Christensson, M. (2016). Biofilm thickness influences biodiversity in nitrifying MBBRs implications on micropollutant removal. *Environ. Sci. Technol.*, 50(17), 9279–9288.

Vigne, E., Choubert, J.M., Canler, J.P., Heduit, A., Sorensen, K.H., Lessard, P. (2011). The role of loading rate, backwashing, water and air velocities in an upflow nitrifying tertiary filter. *Bioresour. Technol.*, 102(2), 904–912.

Waqas, S., Roil Bilad, M., Man, Z., Wibisono, Y., Jaafar, J., Mahlia, T.M.I., Khan, A.L., Aslam, M. (2020). Recent progress in integrated fixed-film activated sludge process for wastewater treatment: A review. *J. Environ. Manage.*, 268, 145–168.

Water Environment Federation (WEF) (2011). *Biofilm Reactors*. McGraw-Hill, New York.

Yerrell, K., Gobbie, M., Dold, P., Jones, R.M., Sickerdick, L. (2001) Full-scale demonstration of a free-moving media IFAS process for enhancing nitrification performance. In *Proceedings of the 74th Annual Water Environment Federation Technical Exposition and Conference*. Water Environment Federation, Alexandria.

Young, B., Delatolla, R., Kennedy, K., Laflamme, E., Stintzi, A. (2017). Low temperature MBBR nitrification: Microbiome analysis. *Water Res.*, 111, 224–233.

3

Trickling Filters and Biofilters

3.1. General information on biofilms

Biofilm formation is a complex process influenced by extracellular proteins and polymers, bacteria concentrations in the surrounding environment and the bacteria forming the biofilm. A bacterial biofilm is described as a community of self-regulated bacteria that have colonized a surface and got structured into a community whose structure consists of a self-generated polymer matrix to facilitate their development in the environment.

Biofilm is used for wastewater treatment, where it is connected to biological oxygen demand (BOD), ammonia, nitrogen, nitrate and dissolved O_2. Wastewater nutrients induce the growth of microorganisms that are used to remove pollutants from wastewaters. Biofilm reactors such as rotating biological contactors, biological aerated filters, fluidized biofilm beds, MBBRs and IFAS reactors are used to remove pollutants from domestic wastewaters as well as a variety of industrial wastewaters. Biofilm reactors can eliminate organic pollutants and nitrogen substances with high kinetics.

Biofilm formation is a dynamic process. Getting attached to the biofilter material, the high concentration of microorganisms increases in time and must be maintained in the system in order to reach the best performances. The concentration of microorganisms forming the biofilm is higher than that of suspended microorganisms. As such, better metabolism and better biodegradation take place in the biofilm compared to those of the system of suspended microorganisms. The formation of glycocalyx (exopolysaccharides) mainly contributes to the structure of microorganisms and to their attachment to the surface. A combination of forces acts during the attachment of microorganisms such as the formation of covalent bonds, the electrostatic interaction and the hydrophobic interaction. The partial covalent bond between the hydroxyl group of the surface and the microorganisms also has its contribution. These forces depend on several factors such as the species present on the material, the surface of the material and the environmental and hydraulic conditions. At the beginning of the adsorption process, electrostatic interaction forces are stronger than the other forces involved in this process. For example, in the case of inorganic materials – active carbon, ceramics, expanded clays – metal hydroxides are formed at their surface. These metal hydroxides are responsible for the attachment reactions facilitated by covalent bonds.

Biofilm evolution involves several stages such as the adhesion of macromolecules (e.g. proteins, polysaccharides, nucleic acids and humic acids) and of smaller molecules (e.g. fatty acids, lipids and pollutants such as polyaromatic hydrocarbons and polyaromatic and polychlorinated biphenyls) on the surfaces via Van der Waals interactions; irreversible adhesion of bacteria via their flagella, fimbriae and pilli; bacterial cell proliferation; and EPS secretion. Cells begin to attach. The adhesion of bacteria to a surface is followed by the production of viscous adhesive substances, extracellular polymeric substances (EPSs). These consist mainly of polysaccharides and proteins.

EPSs are definitely an aid to the formation of mature biofilms by forming a viscous film known as "biofilm matrix". There are three stages: adsorption, consolidation and colonization. Once a mature biofilm is established at the surface, it actively propagates and ends up by covering the entire surface. The mechanisms of propagation in the mature biofilms are more complex than those of the initial attachment. In the mature biofilms, microorganisms are drowned in the EPS layer. Figure 3.1 shows an image of a mature biofilm acquired by scanning electron microscope (SEM), which shows bacteria immersed in an EPS matrix attached to a surface (Aesoy et al. 1998).

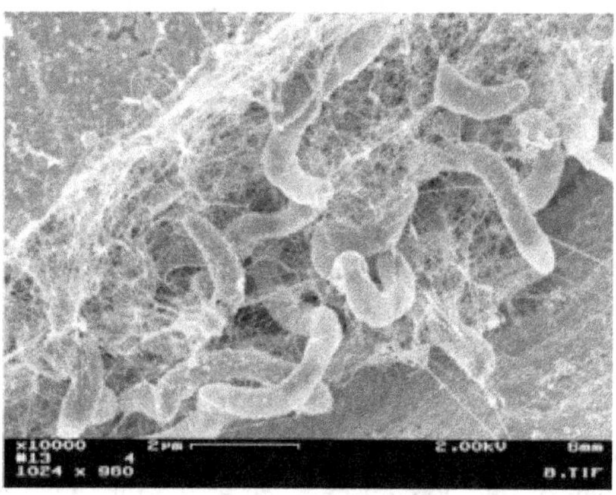

Figure 3.1. *Scanning electron microscope image of a biofilm of Desulfovibrio desulfuricans G20 immersed in EPS (Aesoy 1998)*

In the natural environment, biofilms are formed with various species such as algae, bacteria, fungi and protozoa. Bacteria present in the biofilms can survive under unfavorable environmental conditions such as pH modifications, the presence of toxic substances and free radicals and a low available amount of nutrients. The matrix of exopolysaccharides (EPSs), whose thickness ranges between 0.2 and 1.0 mm, is over 65% of the total biofilm, while microorganisms are the rest of this fraction (Gilles and Bourdon 1987).

These biofilms develop on supports nourished from the organic material and the nutrients contained in the surrounding wastewaters. In these systems, there is a difference between the density gradient between the biomass-coated support and the density of water used inside the reactor that allows establishing a speed gradient between the liquid and the external layer of the biofilm. Consequently, bacteria cells continually exposed to new substrates tend to increase their activity.

Water retention capacity and biofilm mobility are responsible for the diffusion inside the matrix of the biofilm composed of nutrients, polymers and metabolites. The D-glucuronic, D-galacturonic and mannuronic acids contribute to the anionic property of biofilms. Inorganic and organic substances secreted by the microbes provide favorable conditions for the adhesion of microbial cells to the substrate.

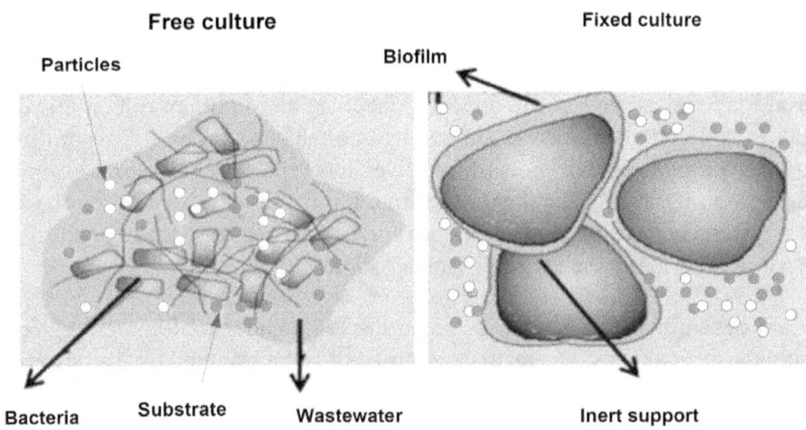

Figure 3.2. *Free culture and fixed culture. For a color version of this figure, see www.iste.co.uk/gaid/wastewaterengineering2.zip*

Several saccharolytic enzymes are produced from bacteria communities of the biofilm. These enzymes are involved in the detachment of microbes from the surfaces to colonize them in a new place. Microorganisms such as bacteria, protozoa, invertebrates and fungi are used to form the biofilm. Heterotrophic bacteria such as *Alcaligenes*, *Bacillus*, *Escherichia coli*, *Nocardia*, *Pseudomonas*, *Streptococcus faecalis*, *Sphaerotilus*, *Thiobacillus* and *Zooglea* predominate in the biofilm. Environmental factors such as bacteria metabolites, the level of oxygen, pH and nutrients influence biofilm formation. A hydrophobic surface, low salinity, low temperature and pH from 7 to 8 are parameters favorable to biofilm formation. Wastewater content of dissolved oxygen, temperature and nutrients determine the bacterial abundance in the biofilm. The pH modifies the bacteria adhesion property by modifying surface electric charges.

Repulsion occurs when the distance between the bacteria and the surface ranges between 10 and 20 nm. Biofilm formation is influenced by EPS concentration. EPSs molecular weight ranges between 500 and 2,000 kDa. They contain glucans, fructans and cellulose derived from saccharose. They are either polyanionic or polycationic. Water (97%) is a key element of biofilms and has a significant impact on the flow of nutrients in the matrix of the biofilm. The biofilm layer has a thickness of about 10 to 30 mm. Biofilm formation occurs under various conditions, but most of them require the use of a humid surface, which can be inert or living. Wastewater facilitates bacteria's adherence to surfaces covered by a layer consisting of polymers. This layer is formed several minutes after the contact and continues to extend for a long period. At this stage of biofilm development, flagellae and pili (structures consisting of fibers) facilitate bacterial adhesion to the surface. Various physical forces, such as Van der Waals bonds, the electrostatic forces and various variables, also play a role in the attachment of bacteria. Flagella, as well as pili and fimbriae, play an important role in this process. Another notable parameter is the surface hydrophobicity. When it is significant, the increased hydrophobicity of the surface drives less repulsion between the surface and the bacteria during the development of the biofilm. Bacterial adherence can be affected by the physical characteristics of the surface such as the porosity, texture and roughness of the surface, besides the chemical characteristics.

When designing biofilm processes, engineers consider biofilms as aggregates of microorganisms immersed in a matrix composed of EPSs excreted by the bacteria fixed on a support.

There are at least four parameters taken into consideration by the designers of biofilm processes:

– the support on which microorganisms are fixed and its characteristics;

– the biofilm (microorganisms and matrix);

– the substrate and the available nutrients;

– the oxygen transfer conditions.

Biofilm reactors used for wastewater treatment are designed to be used by optimizing the factors that manage biofilms and to ensure the best performances of pollutant reduction. The biofilm activity in a reactor can be evaluated based on a mass balance on the substrate (COD or BOD) limiting the growth in the reactor:

$$\text{Biofilm activiy} = \frac{(\text{Lo}-\text{Lf})\text{Q}}{\text{A}}$$

where:

– L is the concentration of the substrate at the inlet and outlet of the reactor ($kg.m^{-3}$);

– Q is the feed flow rate ($m^3.s^{-1}$);

– A is the support surface covered by the biofilm (m^2).

The activity of biofilm at the scale of the reactor is the average flux of substrate crossing the surface of the biofilm. The concentrations of active biomass inside the biofilm are estimated to be between 10 and 60 g of volatile suspended solids per liter of biofilm. This range can be compared to that of reactors with free culture, generally 3 to 8 g of VSS per liter of volume of the reactor. The low value of this range is compared to the activated sludge processes coupled with a clarifier, and the value is compared to membrane bioreactors.

In biofilm reactors, bacteria fixed on supports detach periodically from the matrix and are evacuated from the system with the outlet flow.

Figure 3.3 presents a conceptual illustration of various types of biofilm reactors.

These reactors can be classified depending on whether the biofilm is attached to a fixed support or is mobile inside the reactor:

1) Trickling filters: coarse-sized fixed support or plastic plates. Wastewater flows at the surface of the biofilm in a top to bottom direction and the air flows to the bottom (forced aeration).

2) Immersed biofilter: wastewater flows through the reactor in an upward movement. The support can be fixed and small in size (biocarbon, expanded shale material, Biodagene) or may consist of semi-fixed polystyrene beads (Biostyr®).

3) Fluidized bed: wastewater flows through the reactor in an upwards motion. The support consists of polystyrene balls and is semi-fixed inside the reactor.

4) Biofilter: wastewater flows through the reactor in a downwards motion. The support is fixed and small in size (Biofor®).

5) MBBR: wastewater is mixed in a reactor containing plastic chips. The other fraction of the volume consists of activated sludge. The support is mobile but is maintained in the tank.

6) Hybrid system: the reactor consists of two types of materials of different densities that do not mix. The Biostyr® Duo consists of plastic supports (MBBR type) and polystyrene balls (Biostyr®). Wastewater has an upward motion.

7) Membrane bioreactor: the pollutant is degraded into an activated sludge, and then the water is filtered through a membrane adjoining the biological tank.

There are many biofilm processes (Figure 3.3).

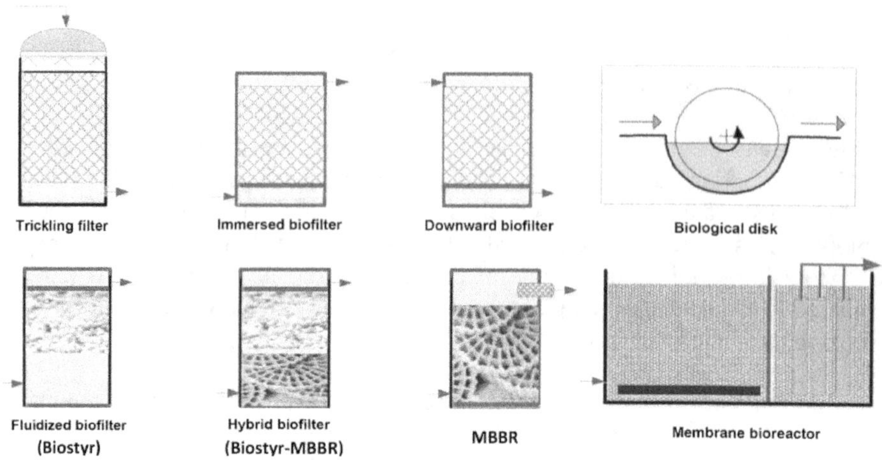

Figure 3.3. *Biofilm processes*

3.2. Trickling filters

The trickling filter is a biofilm reactor on fixed supports. This treatment technique is inspired by ground filtration. It was implemented for the first time at the beginning of the 20th century in Birmingham. It was already known that the self-purifying capacity of soils allowed a biodegradation of organic materials. The objective was to reduce the volume and the surface required for purification by improving aeration in the system. Therefore, the filtering material had to be thicker, thus allowing biodegradation of organic materials inside the filter. A microbial film develops on the inert material, allowing wastewater to flow under gravity. It is important to note that the term "microbial film" has synonyms such as zooglea, bacteria film or biofilm.

Organic materials are associated with oxygen, which plays the role of electron acceptor in the chain of respiratory processes of microorganisms. The degradation of organic materials (BOD) is reflected by the formation of decomposition products such as CO_2 and NO_2, which diffuse in the opposite direction in the flowing liquid and can thus be found in the treated effluent.

3.2.1. *Role of the external support surface for bacteria fixation*

Fixation phenomena have been studied in a very wide range of cases such as adsorption on mineral or organic particles, in marine environments and on dental plaque.

According to Zobell (1943), who conducted microscopic studies on seawater samples, many bacteria are fixed on particles, while very few are free in water. Plankton is also a support for microorganisms, as it is a significant source of energy and nourishment. Moreover, its often viscous consistency facilitates the adhesion of bacteria.

Bacteria growth is often related to the surface they adhere to. Bacteria growth seems to be related to the support characteristics, and their distribution is not uniform at the surface of solids. They agglutinate in locations where substrate adsorption is most probable. Bacteria activity is proportional to the concentration of nutrients and any factor tending to concentrate organic materials will have a positive effect on bacteria activity.

Since 1963, Bennett called this coating "glycocalyx"; it has chemical properties allowing it to fix certain substances (Bennett 1963). According to Wilkinson (1958), glycocalyx may act as an ion exchange resin; other authors focused on determining the protective role it has for bacteria, in relation to inhibitory substances such as amoeba, phagocytes, bacteriophages and endotoxins. There is a wide range of polysaccharides according to their nature. Chains are linear or ramified (branched), containing aldoses, ketoses, uronic acids and aminosugars, with sugar content varying depending on the type of bacteria. For example, for Gram-negative bacteria, we generally obtain glucose, rhamnose and galactose, while for Gram-positive bacteria, we can note fucose, hexoses, glucose and various (amino uronic, hexuronic, glucuronic) acids.

3.2.2. *Ecology of trickling filters*

The microscopic observation of the microbial film present on the surface of the material reveals a very rich and diversified bacteria population.

Bacteria clusters can be noted, essentially composed of heterotrophic Gram-negative bacteria such as *Pseudomonas*, *Achromobacters*, *Alcaligenes* and *Flavobacterium*. Filamentous bacteria (*Zooglea* among others) are also observed.

Fungal mycelia such as *Sepedonium*, *Ascoidea*, *Fusarium* and *Geotrichum* can also be found. In uncovered trickling filter areas where there is more light, single cell or filamentous algae can be found. They may involve, for example, *Chlorella*, *Chlorococcum* and *Phormidium*.

3.2.3. *Creation of trickling filters*

3.2.3.1. *Operation*

A trickling filter consists of a permeable medium made from a bed of rocks, scoriae or plastic on which wastewaters are distributed to flow, as shown in Figure 3.4.

Wastewaters enter the bioreactor through a distribution system, flow to the bottom on the biofilm surface and the air goes up or down through the material, where it diffuses through the flowing liquid and in the biofilm.

The equipment of the trickling filters normally includes water distribution, a concrete structure, a rock or plastic support material and a drainage and ventilation system. A clarification-based liquid–solid separation is required, and it is generally made with circular or rectangular clarifiers. Rotary screens can be used.

Along the water stream, there is pretreatment for the removal of waste (screening), sands (grit removal), oil and grease (oil removal/degreasing).

Wastewater then flows into a primary settling tank to remove a part of the suspended solids, particulate BOD_5 and particulate COD, as well as all the forms of mineral particles that settle under the operating conditions of the work.

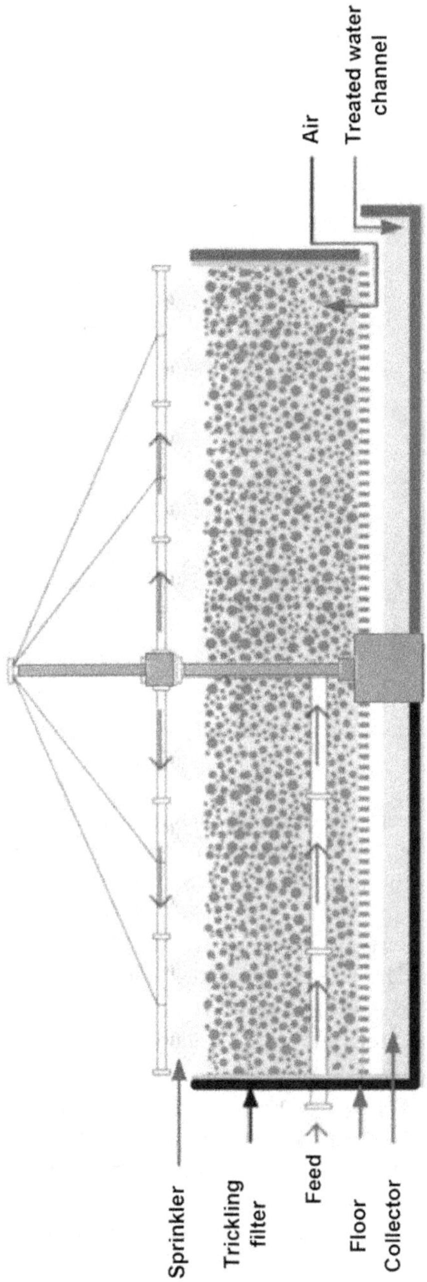

Figure 3.4. *Schematic representation of trickling filter*

There are essentially two systems for effluent distribution on the materials forming the trickling filter: fixed nozzle distributors and rotary distributors. Because of poor efficiency, fixed nozzle distribution is gradually being abandoned. Rotary distributors can be hydraulically or electrically driven. The properly designed distribution system allows efficient wetting of materials, fixing the biofilm and a sequential feed. Poor wetting of the support may lead to dry areas, inefficient treatments and sources of odors. An electrically or hydraulically driven rotary distributor controls the rotational speed independently of the ingoing wastewater flow rate and can maintain the expected hydraulic flow rate (Metcalf & Eddy Inc. 2002).

a) b)

Figure 3.5. *Sprinkler with: a) longitudinal distribution and b) circular distribution*

Rock or scoriae beds can reach 60 meters in diameter and up to 10 m height with sizes of rocks ranging between 25 and 102 mm. Many rock media provide about 149 $m^2.m^{-3}$ and about 40% of the void index.

Trickling filters with plastic materials have a smaller diameter (6–12 m) and a smaller span (4–12 m height). These filters are rather similar to towers, with materials arranged according to various configurations (e.g. vertical flux, transverse flux, cross flux, etc.). Cross-flux materials offer a better flow and good efficiency, particularly at low organic load. A part of the settled effluent is recirculated to improve wetting and facilitate excessive sludge removal. Wastewaters pass through the filtering material on which the biofilm develops. In this biofilm, microorganisms oxidize the organic material of wastewaters into carbon dioxide while generating a new biomass. To keep microorganisms alive, the filtering medium must be maintained humid; consequently, effluent recirculation is applied to prevent the desiccation of the biofilm layer under low flow rate conditions. It is therefore essential to have sufficient air available to provide O_2 to microorganisms composing the biofilm and avoid anaerobic conditions that would disrupt bed performances. Air is brought to the filtering bed by natural ventilation. If the temperature of ingoing wastewaters is below the ambient temperature, air flows to the bottom through the trickling filter. Conversely, air will circulate to the top if wastewater temperature exceeds the ambient temperature.

The underground drainage system collects the treated effluents and the removed biomass in a percolating filter. Moreover, it allows air circulation to ensure proper operation. The drainage system under the floor is designed to carry a maximal wastewater flow rate for which the percolating filter is designed.

Biofilm thickness may reach up to 10 mm, depending on the operating conditions. As the biofilm thickness increases, anaerobic conditions are generated at the surface of the filter. At this stage, the biofilm enters an endogenous respiration phase given the limited intake of organic material in the inner depths of the biofilm. The microorganisms can no longer attach to the filtering medium, which leads to the detachment of the biofilm under the effect of gravity and of ingoing wastewaters. This is referred to as desquamation.

Trickling filters are generally circular with diameters ranging from several meters to several dozen meters.

3.2.3.2. *Support materials*

Trickling filters are made based on several parameters:

– choice of material;

– effluent distribution;

– recycling of treated water.

Support materials must be locally available and sustainable; have a specific surface, a high percentage of void space to allow a proper air and water circulation and a resistance to attrition; and must be insoluble in wastewaters. Their shape must be round or cubic. Only 5% of the media must have the longest dimensions, higher than three times the smallest dimensions:

– support size: 25–75 mm;

– specific surface: 45–60 $m^2.m^{-3}$ (rocks);

– specific surface: 90–150 $m^2.m^{-3}$ (plastic materials).

Various materials meeting these conditions are used. The following materials are used as media support for trickling filters: pouzzolane, clinker, coke, pumice and also plastic materials.

The generally accepted granulometry varies depending on the country. In the United States, it ranges between 40 and 80 mm, while in Great Britain, a range between 20 and 50 mm is preferred. In India, the recommended support size varies between 25 and 75 mm. Indeed, the choice of granulometry involves two contradictory demands. The material must be quite significant to avoid rapid clogging, and it also has to be quite fine to allow the largest possible contact surface. For example, a grain of spherical shape with a diameter ranging between 40 and 80 mm ensures a contact surface of 95 $m^2.m^{-3}$, while a grain with a diameter ranging between 20 and 40 mm has a contact surface of 190 $m^2.m^{-3}$. Assuming that the biofilm thickness is identical in both cases, it is clear that the choice of the second range of dimensions (20-40 mm) will allow the purification of twice as much wastewater.

This does not, however, exclude the possibility of faster clogging. The bed height is then modified by adopting lower dimensions when granulometry is more significant.

In small facilities, the media support of trickling filters often consists of layers of various granulometry, the coarsest materials being laid in the lower part of the filter to allow more consistent ventilation. Proposed materials include PVC plastic materials, Cloisonyl tubes or still waffled plates in stacked blocks (Surfpac, Flocor, etc.).

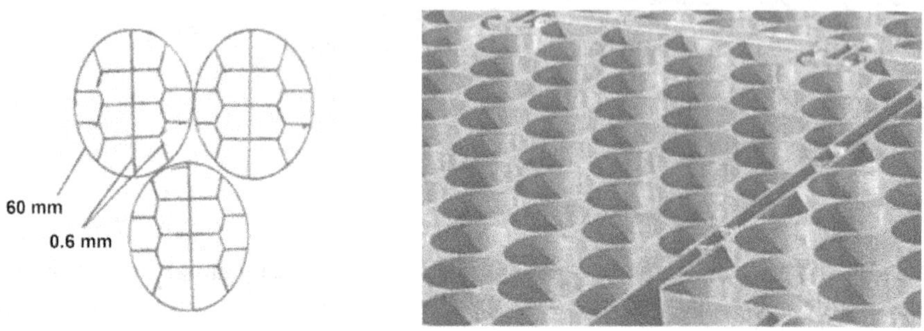

Figure 3.6. *Cloisonyl tubes (seen from above)*

Due to the high specific surface areas and the significant void ratios of support materials, they appear as an interesting material to the extent that they allow the consideration of significant heights and therefore ground surface reduction.

The use of plastic trickling filters finds its application particularly in the following cases:

– biological pretreatment in facilities with high BOD as industrial discharges;

– at the head of an already existing municipal or industrial station that is overloaded;

– before the discharge in a sewer network in order to reduce the purifying workload and remove a part of the toxic substances in case of potential accidents.

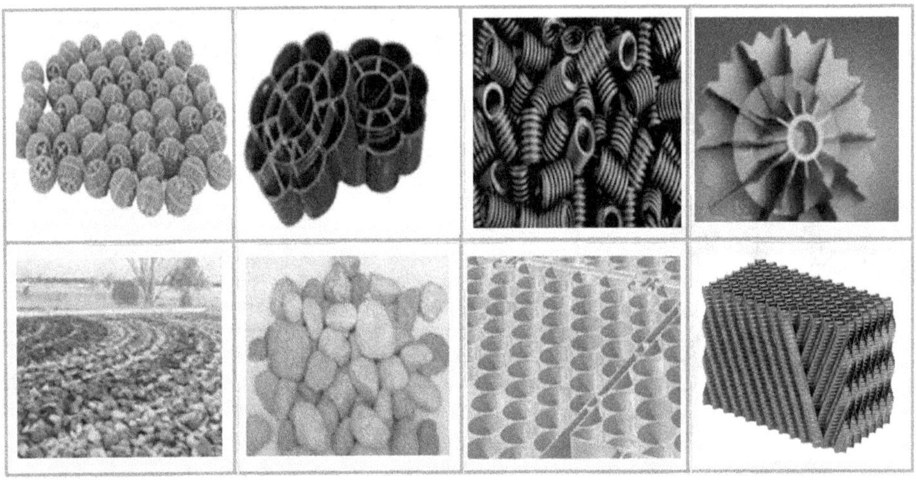

Figure 3.7. *Presentation of various rock and plastic materials*

The types of trickling filter media include rocky materials and plastic materials with vertical or transversal flux, manufactured with smooth or corrugated sheets.

Plastic materials are almost exclusively used for new trickling filters or during their rehabilitation.

However, trickling filters consisting of rocky stones continue being used and have the expected performances if they are properly dimensioned and designed.

Type	Material	Dimensions (mm)	Density ($kg.m^{-3}$)	Specific surface ($m^2.m^{-3}$)	Void ratio (%)
Rocks					
Pebbles	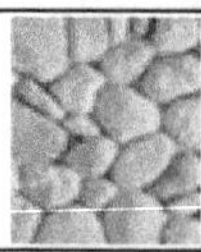	24–76	442	62	50
Gravel	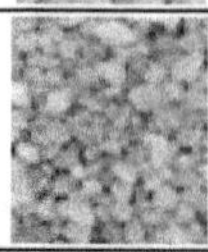	76–128	1600	46	60
Volcanic rocks					
Pouzzolane	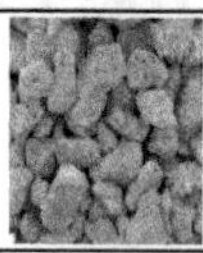	24–75	1,300	100	45
Plastic materials					
Cloisonyl	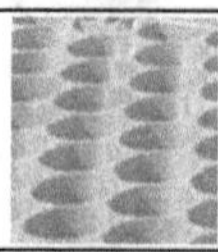	–	80	220	94
Standard Surfpac		–	65	80	94
Flocor		–	38	80	98
Tangential		0.61 × 0.61 × 1.22	24–45	100–138–223	95
Vertical		0.61 × 0.61 × 1.22	24–45	102–131	95
Random		0.185 × 0.051	27	98	95

Table 3.1. *Characteristics of several support materials of trickling filters (Lewandowski and Boltz 2011)*

The higher specific surface and the void space in the plastic modules offer higher hydraulic loading rates, improve the oxygen transfer and allow easier control of the thickness of the biofilm. Rocky materials have various dimensions even if the "ideal" size would be 50 mm. Given the structural demands associated with their unit weight, trickling filters with rocky media are shallow compared to those involving plastic media. The rocky material generally has a small specific surface, a low void ratio and a high weight. Despite the strong recirculation of waters upstream of the units, the low void ratio limits hydraulic loading rates, since if this parameter is excessive, it may lead to the formation of puddles, limited transfer of oxygen and poor performance of the biological filter. The performances of the trickling filters with rocky materials can be improved by installing mechanical ventilation, replacing the natural ventilation. Plastic materials have limited clogging risks, high hydraulic loads due to their significant specific surface and a void ratio above 94% that may reach 98%.

Table 3.1 presents the characteristics of rocky materials and of plastic materials.

The two types of materials also remove BOD_5. However, this can only be achieved by operating with lower organic and hydraulic loading rates with rocky materials. Concerning nitrification, many studies have shown that it is better with rocky materials for which the small size and the loads applied facilitate the presence of nitrifying bacteria and their growth kinetics.

According to researchers, lower rates are reached with plastic media because of sequential watering that induces less efficient wetting. Vertically oriented plastic materials are generally accepted as being perfectly adapted to high organic loading rates.

In some cases, the rocky material is placed in the upper layer to improve the distribution of wastewaters and the plastic material fills the remaining volume (Lewandowski and Boltz 2011).

As a general rule, the upper layer of the plastic support is covered by a grid to prevent its degradation by UV radiation. Not all materials are self-supporting and they then require a floor to keep them in the biological filter.

3.2.3.3. *Wastewater feed distribution*

The wastewater distribution system is very important for the quality of purification. The first distributors (sprinklers) were fixed. They consisted of gutters feeding perforated cylindrical tubes. They required a loss of load from 1 to 2 m to continuously feed the structure. They also created preferred paths for the passage of wastewater through the filter. The total volume of the trickling filter was not fully used, disrupting the purification quality. Filter feeding was improved by equipping them with rotary distributors (or sprinklers).

Figure 3.8. *Rotary distributors*

Rotary distributors are represented by hydraulic tourniquets with two, four or six horizontal arms, allowing a homogeneous distribution of wastewater on the bed surface, all the more so as the distribution of the holes of distributors present on the arm is calculated to ensure equal sprinkling for larger size devices.

Arms are normally set into motion by the water load, but it is often preferred to attach an electric motor whose low energy consumption renders the system advantageous from both a technical and economic perspective.

In the case of rectangular trickling filters, the feeding device consists of mobile ramps or trolleys with an alternating motion driving a system of perforated pipes distributing wastewater. The load loss in this type of distribution varies between 1 and 2 m.

3.2.3.4. *Water recirculation*

Effluent recirculation is often necessary, and in some cases, it is even compulsory, as it offers several advantages. It involves the filter self-cleaning by maintaining a sufficiently significant flow rate to erode the bacteria film accumulated on the material.

It allows the improvement of the purifying efficiency, to the extent that the dilution of the ingoing effluent reduces the concentration of BOD to be treated and allows the discharge of higher quality water. It facilitates permanent seeding of the bed and brings in oxygen and nitrates, allowing the activation of the purifying work.

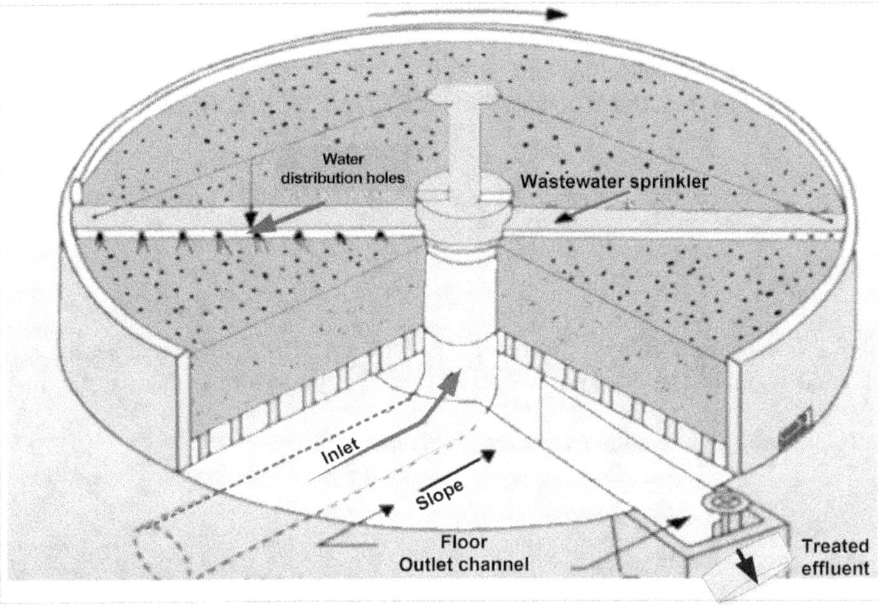

Figure 3.9. *Hydraulic operation of the trickling filter*

3.2.3.5. *Implementation of trickling filters*

Implementing trickling filters is relatively simple, as in fact, a trickling filter is considered a pile of rocks equipped with a device for effluent distribution. As a first stage, the floor that must support the materials is ordered. It is arranged so that the effluent trickling on the bed flows easily without leaving sludge deposits that would rapidly clog the bed. The Imhoff company recommends the implementation of a double floor, with a first floor serving as support and the role of the second one being to remove the treated effluent. The existing air mattress allows for bed ventilation, thus facilitating the oxygen contribution for the biomass. The distance between floors must be sufficient to prevent the accumulation of sands, fragments of crumbling materials and sludge so that the natural aeration of the bed is not disrupted. The lateral walls of the trickling filters often consist of concrete walls. However, for economic reasons, one can also use masonry elements such as bricks or metallic gabions.

Depending on the adopted geometric shape, we then choose the effluent distribution device. When trickling filters are intended for treating wastewaters of a small agglomeration, the height of the material is small and allows natural aeration. This occurs vertically in one direction or another depending on whether the density of the air present in the bed is higher or lower than that of the external air. Assuming that the temperature inside the bed is equal to that of wastewaters, and that in winter its value is 13°C, it is obvious that during this period, external air being colder (higher density), air will circulate from bottom to top. On the contrary, in a warmer season, ventilation will be from top to bottom. This is also true for two different temperatures during the same day.

The pressure resulting from the difference in temperature is given by:

$$P = 353 \left(\frac{1}{Tfr} - \frac{1}{Tch} \right) H$$

where P is expressed in mm of water, Tfr is the temperature in the cold season (°K), Tch is the temperature in the hot season (K) and H is the height of the material (m).

The average temperature of the year is given by:

$$Taverage = \left(\frac{Tfr - Tch}{Ln(\frac{Tfr}{Tch})} \right)$$

where temperatures are expressed in K.

Natural ventilation is normally used for heights of the bed up to 4 m. When bed dimensioning requires a more significant height, natural ventilation is no longer sufficient, as the head loss is higher, caused by more significant sludge content inside the bed. Forced ventilation is then established. This is readily conducted by covering the bed and setting a ventilator at the top whose power is calculated to allow sufficient oxygen intake.

For forced aeration, it is recommended to implement a flow rate of 18 m^3 of air.m^{-2} of surface.h^{-1}.

The O_2 needs are given by the following relations:

– carbon removal:

QO_2 (kg O_2/kg applied BOD) = (20 kg/kg) [0.80 e^{-9Lo} + 1.2 $e^{-0.17Lo}$] Cp

– carbon removal and nitrification:

QO_2 (kg O_2.kg^{-1} applied BOD) = (40 kg/kg) [0.80 e^{-9Lo} + 1.2 $e^{-0.17}$ Lo + 4.6 Nox/DBO] Cp with

- Lo being the BOD input in kg BOD.m^{-3}.d^{-1};
- Nox/BOD being the ratio of N oxidized input/BOD input (mg.mg^{-1});
- Cp being the peak coefficient, related to the maximal organic load applied;
- air having the molecular weight 28.97;
- density at 0°C and 1 atm = 1.2928.

The following relation is used to determine air density at a given temperature T°C:

$$\rho\text{air} = \left(\frac{\text{PM}}{\text{RT}}\right)$$

with:

– P being the atmospheric pressure = 1.01325.10^5 N.m^{-2};

– M being the mole of air = 28.97 kg.kg^{-1}.mole^{-1};

– R being the universal gas constant: 8,314 N.m.kg^{-1}.mole^{-1}.K;

– T being the temperature K = 273.15 + T °C.

For example, at 10°C, air density is:

$$\rho\text{air (10°C)} = \left(\frac{1.01325\ 10^5 \text{ x } 28.97}{8{,}314 \text{ x } (273.15 + 10)}\right) = 1.246 \text{ kg.m}^{-3}$$

The change in atmospheric pressure between two altitudes a and b is given by:

$$\left(\frac{\text{Pb}}{\text{Pa}}\right) = \exp\left(-\frac{\text{gM(Hb} - \text{Ha)}}{\text{RT}}\right)$$

with:

– g being the gravitational acceleration = 9.81 m.s^{-2};

– M being the mole of air = 28.97 kg.kg^{-1}.mole^{-1};

– H being the elevation in m;

– R being the universal gas constant = 8.314 N.m.kg^{-1}.mole^{-1}.K;

– T being the temperature K = 273.15 + T °C.

The percentage of O_2 in the air (in weight) is 23.18%.

The volume of O_2 per kg of air is: (1/(1.204 (kg.m^{-3})(0.2318)) = 3.58 m^3.kg^{-1}.

The air flow rate at 20°C and 1 atm (in m^3.min^{-1}) is:

$$\text{Q air (m3. min}^{-1}) = \frac{\text{QO2 Qwater Lo (3.58 m3.kg}-1)}{10^3\left(\frac{g}{kg}\right)\left(1{,}440\frac{min}{d}\right)}$$

The air flow rate is corrected depending on T and on pressure according to the ideal gas law:

$$\text{Qair Tamb(m}^3\text{.min}^{-1}) = \text{Qair20°C}\frac{273.15+\text{Tair}}{273.15}\frac{760}{P}$$

with:

– Qair T being the air flow rate at the temperature T °C (m^3.mn^{-1});

– Tair being the temperature of ambient air in °C;

– P being the on-site pressure in mm Hg.

When the temperature is above 20°C, the following relation is recommended:

$$\text{Qair (T° > 20°C)(m}^3\text{.min}^{-1}) = \text{QairTair}\left(1 + \frac{\text{Tair}-20}{100}\right)$$

The head loss (pressure drop) related to the material of the trickling filter is linked to the air flow velocity by the following relation (case of forced ventilation):

$$\Delta P = Np\frac{v^2}{2g}$$

with:

– ΔP being the head loss in kPa;

– g being the gravitational acceleration = 9.81 m.s^{-2};

– v being the surface velocity = Q/S m.s^{-1};

– Np being the resistance factor related to the material.

This factor is given by:

$$Np = 10.33\ H\ e^{(1.36\ 10\text{-}3)/(Q/S)}$$

with:

– H being the height of the material (m);

– Q being the water flow rate ($kg.h^{-1}$);

– S being the passage surface (m^2).

As this factor was developed by Dow for vertical plastic materials and there are no other elements for other materials, a correction factor was proposed by WEF (2000).

Material	Specific surface ($m^2.m^{-3}$)	Correction factor
Rocks, gravel	45	2.0
Vertical flux plastics	100	1.3
Transversal flux plastics	140	1.6
Random plastics	100	1.6

Table 3.2. *Correction factors for head loss calculation*

EXAMPLE.– Calculation of air needs for a trickling filter and of head loss in case of forced ventilation.

– Wastewater flow rate: 10,000 $m^3.d^{-1}$ = 115 $L.s^{-1}$.

– Input BOD_5: 250 $mg.L^{-1}$.

– Water temperature: 12°C.

– Air temperature: 5–30°C.

– Peak coefficient: 1.4.

– Material: transverse plastics.

– Correction factor: 1.3.

– Head loss for valves-pipes: 1.5.

– Bed diameter: 20 m.

– Height of material: 5 m.

– Volume of material: 1,570 m^3.

– Applied BOD_5 loading: 10,000 x 0.25/1 570 = 1.60 kg $BOD_5.m^{-3}.d^{-1}$.

– O_2 needs:

$$QO_2\ (kg\ O_2.kg^{-1}\ applied\ BOD) = (20\ kg/kg)\ [0.80\ e^{-9Lo} + 1.2\ e^{-0.17Lo}]\ Cp$$

$$QO_2\ (kg\ O_2.kg^{-1}\ applied\ BOD) = 20\ [0.80\ e^{-9 \times 1.6} + 1.2\ e^{-0.17 \times 1.6}]1.4 = 25.53$$

$$\text{Qair}\ (\text{m}^3.\text{min}^{-1}) = \frac{\text{QO2 Qwater Lo}\ (3.58\ \text{m}^3.\text{kg}^{-1})}{10^3\ \left(\frac{\text{g}}{\text{kg}}\right)\left(1{,}440\,\frac{\text{min}}{\text{d}}\right)}$$

$$\text{Qair}\ (\text{m}^3.\text{min}^{-1}) = \frac{25.53 \times 10{,}000 \times 250 \times 3.58}{10^3 \times 1{,}440} = 158$$

– Temperature and pressure correction. As T °C > 20°C, we have:

$$\text{Qair Tamb}(\text{m}^3.\text{min}^{-1}) = \text{Qair}20°\text{C}\frac{273.15 + \text{Tair}}{273.15} = 158\frac{273.15 + 30}{273.15}$$

$$\text{Qair Tamb}(\text{m}^3.\text{min}^{-1}) = 175.35$$

For the highest temperature (30°C), we obtain:

$$\text{Qair}(\text{T}° > 20°\text{C})(\text{m}^3.\text{min}^{-1}) = \text{QairTair}(1 + \frac{\text{Tair} - 20}{100})$$

$$\text{Qair}(\text{T}° > 20°\text{C})(\text{m}^3.\text{min}^{-1}) = 175.35\left(1 + \frac{30 - 20}{100}\right) = 192.8$$

3.2.4. *Classification of trickling filters*

Trickling filters are defined by their organic loading rate, which is the daily quantity of organic materials admitted per m^3 of materials. They can be classified into low-loading-rate filters, average-loading-rate filters and high-loading-rate filters. The operating principle is presented in Figure 3.10.

The following can be identified:

– low-loading-rate filters, up to 0.4 kg $BOD_5.m^{-3}.d^{-1}$;

– average-loading-rate filters, up to 0.8 kg $BOD_5.m^{-3}.d^{-1}$;

– high-loading-rate filters, above 0.8 $BOD_5.m^{-3}.d^{-1}$.

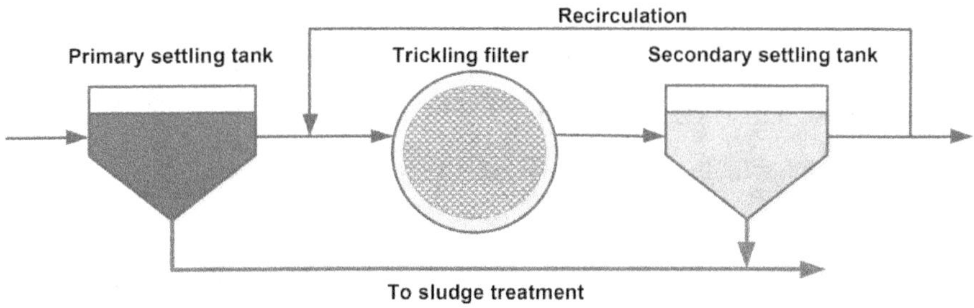

Figure 3.10. *Operating principle of a treatment plant with a trickling filter*

Some authors rightfully admit a fourth class that is related to very high loads (1.5 $BOD_5.m^{-3}.d^{-1}$) appropriate to plastic materials.

3.2.4.1. *Low organic loading rate*

In this type of filter, the contribution of organic materials is low. Low-rate filters are commonly used for organic loads below 40 kg of $BOD_5.m^3.d^{-1}$. In this case, the bacteria present on the material are close to the endogenous state. They mineralize progressively and clog the filter if they are not driven at a high flow rate.

As the feed flow rate is low in this type of facility, it cannot ensure the cleaning of materials. Practicing strong intermittent recycling is required to prevent frequent clogging. Purification efficiencies are about 90% for BOD. The BOD_5 of the treated effluent varies between 20 and 30 $mg.L^{-1}$. With rocky material filters, the limits of hydraulic loading rates vary between 1 and 5 $m^3.m^{-2}.d^{-1}$. These systems have fewer problems than the other types of filters concerning the presence of flies, odors and clogging. Most filters operating under low loading rates are circular and driven by rotary distributors. The height of the material varies between 0.9 and 2.4 meters. A minimum wetting rate of 0.4 liters per square meter and per second is maintained in plastic beds to prevent the desiccation of the material.

Solid materials coming from a low-load filter are in general properly digested, and they consequently produce fewer solids than those from average- and high-loading-rate filters.

It is important to note that a hydraulic loading rate is defined as being the daily volume of water brought per square meter of surface. The dimensioning of these filters (diameter, height) is deduced from the conditions of organic loading rate applied to the material.

EXAMPLE.–

– Connected population: 10,000 inhabitants.

– Daily consumption: 150 $L.hab^{-1}.d^{-1}$.

– BOD_5 per inhabitant: 54 $g.hab^{-1}. d^{-1}$.

– Peak coefficient Cp: 1.4.

– Wastewater flow rate Q average = 10,000 x 150/1,000 = 1,500 $m^3. d^{-1}$.

Assuming that primary settling removes 35% of total BOD_5, the organic load entering the trickling filter is:

$$\text{organic loading rate} = 10{,}000 \times (54 \times (1 - 0.35))/1{,}000 = 351 \text{ kg } DBO_5.\ d^{-1}$$

$$\text{Lo, } BOD_5 \text{ concentration} = 351/1\ 500 = 234 \text{ mg.L}^{-1}$$

By adopting a low loading rate equal to 0.2 kg $BOD_5.m^{-3}.d^{-1}$, the volume of the filter is:

$$V = 351 \text{ kg } DBO_5.d^{-1}/0.2 \text{ kg } DBO_5.m^{-3}.\ d^{-1} = 1{,}755 \text{ m}^3$$

The hydraulic loading rate acceptable for low-loading rate filters ranges between 1 and 5 $m^3.m^{-2}.d^{-1}$. By adopting an average loading rate of 3 $m^3.m^{-2}.d^{-1}$, the total surface of the filter is:

$$S = 1{,}500/3 = 500 \text{ m}^2$$

Let us consider two units with a unit diameter equal to 17.9 m and unit surface of 251.5 m^2.

The unit height of the material is therefore:

$$1{,}755/(2 \times 251.5) = 3.5 \text{ m}$$

While for all trickling filters the presence of a primary settling tank is essential, as it avoids and protects the bed from rapid clogging, the use of a secondary settling tank is not always justified in the case of very-low-loading-rate filters. Collected sludge is dense, non-putrescible and readily settleable.

3.2.4.2. *Average organic loading rate filters*

Average organic loading rate filters can be fed up to 64 kg $BOD_5.m^{-3}.d^{-1}$.

3.2.4.3. *High organic loading rate filters*

Unlike the previous cases, the material is only coated with a thin layer of microbial film, the feed flow rate being sufficiently high to ensure the self-cleaning of materials by adopting adequate recycling rates. The treatment efficiencies are about 60% without recycling and 80 to 90% with recycling. The admissible hydraulic loading rate is about 0.8 $m^3.m^{-2}.h^{-1}$ or about 20 $m^3.m^{-2}.d^{-1}$. The residence time is about 90 mn at peak flow rate. Removed sludge is granular and very putrid. The trickling filter must be followed by a secondary settling tank for which the calculated residence time is identical to that of the primary settling tank. The applied organic loading rates may rise up to 160 kg of $BOD_5.m^{-3}.d^{-1}$. Acceptable hydraulic loading rates are up to 30 $m^3.m^{-2}.d^{-1}$.

For a height of 3 m, natural aeration is often sufficient, and the surface is evaluated at about 1 m^2 for 75 inhabitants. In order to increase bed efficiency, a recirculation of a part of the treated effluent is established. The aim is to ensure three objectives are met.

3.2.4.3.1. Trickling filter self-cleaning

To ensure efficient self-cleaning of the trickling filter, the hydraulic rate must be of the order of 0.8 to 1 $m.h^{-1}$. This value can only be reached by reinjecting a part of the effluent upstream of the filter so that:

$$C_h = (1 + R)\ Qm/S$$

where:

- R is the expected recirculation rate for reaching the load C_h;
- Qm represents the average flow rate of effluent at the station inlet;
- S is the trickling filter surface.

Or:

$$R = (C_h\ S\text{-}Qm)/Qm$$

This recirculation rate should meet the self-cleaning conditions. It is defined as being the ratio of the mixture flow rate and the peak flow rate in dry weather.

For example, for a population of 10,000 inhabitants (daily consumption of 150 liters per inhabitant and per day), the average flow rate is 62.5 $m^3.h^{-1}$. For a value of C_h of 1.2 $m^3.m^{-2}.h^{-1}$, and a surface of 250 m^2, the recirculation rate is:

$$R = (1.2 \times 250 - 62.5)100/62.5 = 380\%$$

3.2.4.3.2. Performance improvement

The average efficiency of a trickling filter operating at high loading rates is about 60%. This does not allow meeting the effluent standards. Indeed, by admitting that the BOD_5 of the settled effluent is 230 $mg.L^{-1}$, an efficiency of 60% leads to an output BOD_5 of 92 $mg.L^{-1}$.

Due to the effluent recycling, the input BOD_5 can be diluted and the expected efficiency can be reached. Taking into account the expected efficiency, the recycling rate can be obtained as follows: for a value of the output BOD_5 equal to 30 $mg.L^{-1}$, a balance can be written reporting the effluent BOD_5 and the mixture BOD_5.

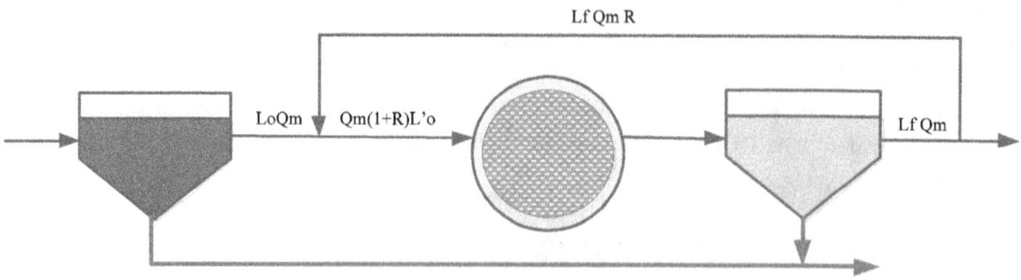

Figure 3.11. *Recirculation of the treated effluent*

The following can be written:

$$(1 + R)\ Qm\ L'o = Lo\ Qm + Lf\ Qm\ R$$

or:

$$R = (Lo - L'o)/(L'o - Lf)$$

where Lo is the BOD_5 of the input effluent, Lf is the BOD_5 of the treated effluent, L'o is the BOD_5 of the mixture and Qm is the average flow rate of the raw effluent.

By considering a peak flow rate, the recycling rate can be established on the same balance:

$$(1 + R)\ Qp\ L'o = Lo\ Qp + Lf\ Qp\ R$$

$$R = (Lo - L'o)/(L'o - Lf)$$

According to the previous example, the peak flow rate is equal to 1.5, and we obtain:

R = (1.2 x 250 – 62.5x1.5)100/(62.5 x 1.5) = 220%

L'o = (Lo Qp + Lf Qp R)/(1 + R) Qp

L'o = (230 x 67.35 x 1.5 + 30 x 67.5 x 1.5 x 2.2)/[(1 + 2.2) 67.5 x 1.5] = 92.5 $mg.L^{-1}$

3.2.4.3.3. Seeding the material

Recycling allows maintaining constant humidity on the material and a permanent seeding of the trickling filter, especially as this is reflected by a deeper penetration of the bacterial flora inside the filter, and a fortiori this influences the efficiency of the treatment.

3.2.5. *Implementation of recirculation*

3.2.5.1. *Trickling filter with a single stage*

Various methods are possible. In all cases, on the one hand, bed clogging is prevented and, on the other hand, constraints due to discharge conditions can be taken into account. Various possible diagrams are presented in Figure 3.12.

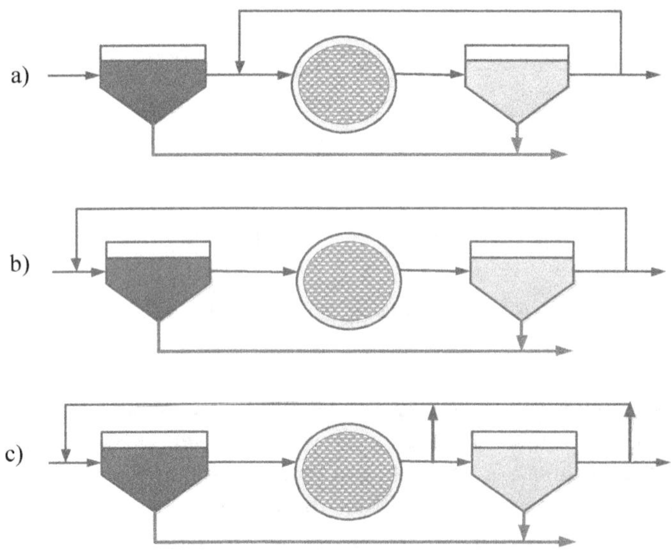

Figure 3.12. *Several possible recirculation diagrams*

In (a), the settled effluent is recycled at the head of the trickling filter.

In (b), the effluent is recycled at the head of the primary settling tank. In this case, this work will be calculated to receive a flow rate Q(1 + R.). The carried sludge is retained in the primary settling tank, which is consequently dimensioned.

In (c), the effluent can be recycled at the head of the primary settling tank and at the head of the trickling filter. The primary settling tank will be consequently dimensioned. A part of the sludge is generally recycled in the primary settling tank. The purpose is to facilitate and accelerate sedimentation. There are times when the entire sludge recovered in the secondary settling tank is recycled, and primary and secondary sludge is gathered and treated.

3.2.5.2. *Configuration of serial trickling filters*

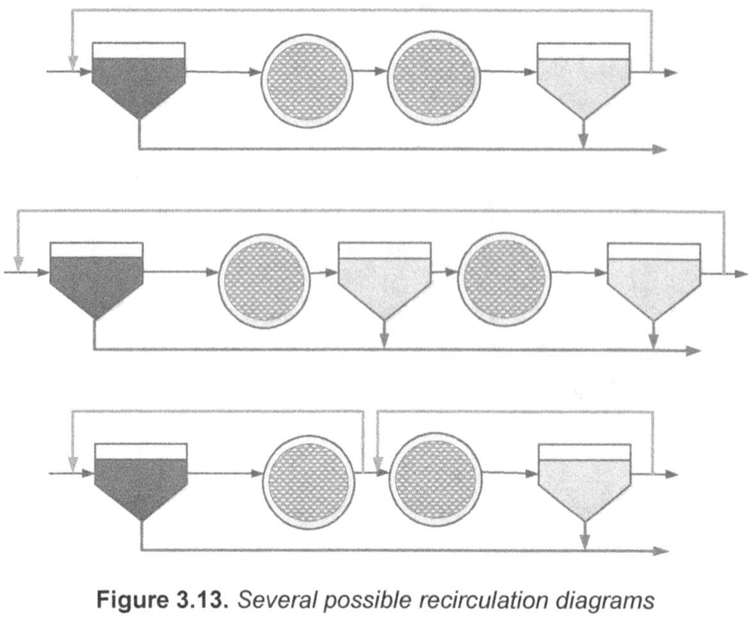

Figure 3.13. *Several possible recirculation diagrams*

In some cases, to improve the quality of the effluent, two serial trickling filters are used. This solution involves treating wastewaters by successive passage in two filters, so that the treatment is equivalent to that of a single filter having a surface and a height corresponding to the sum of the dimensions of the two filters. However, having two filters of different dimensions is possible. The separation in two stages has the advantage of allowing better aeration, essentially due to the small height of each of the two filters. The flow diagram involves a primary settling tank, with the two serial beds having the same secondary settling tank. The clogging of the second trickling filter that is fed by the unsettled effluent of the first bed must be frequently controlled. An alternative is to design the second trickling filter with a different material, accepting higher passage speed, allowing the prevention of clogging and ensuring permanent self-cleaning.

A second settling tank is sometimes inserted between beds. This proves to be more costly for the investment. There is a very wide range of recycling possibilities, which have the advantage of offering several combinations. The treatment is chosen based on economic imperatives and in compliance with removal standards.

3.2.6. *Using trickling filters*

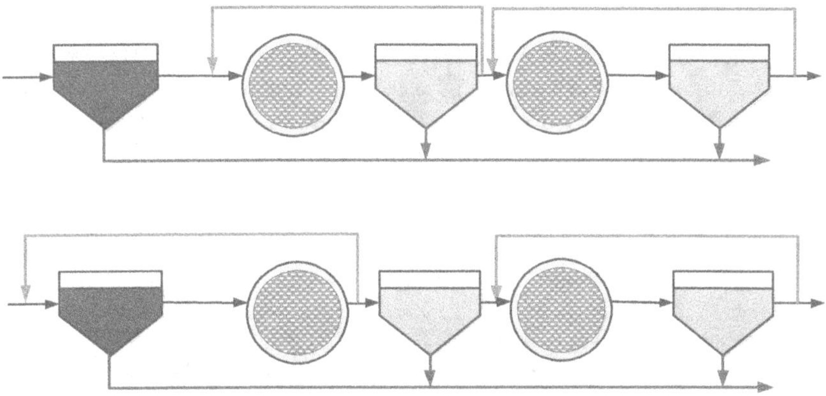

Figure 3.14. *Various configurations of serial trickling filters*

Trickling filters are well adapted to any treatment possibility to the extent that the effluent to be treated contains biodegradable organic materials. Unlike activated sludge, they support overloading effects that may occur in the unit network. They also support toxic shocks (especially metallic ions) through the adsorption of these elements on the bacteria film.

Bacteria adhesion is reflected by the formation of hairy fibers known as “glycocalyx” that serve as a passageway and also as a trap in relation to mineral and organic materials. This is why toxic products are found blocked on the external surface of the cell membrane.

In a treatment plant treating a mixture of urban and industrial wastewaters, the analysis of sludge collected in the secondary settling tank led to ion values of about 5 to 20 $mg.g^{-1}$ of VSS. Despite these high values, filter operation was satisfactory.

Trickling filters are, however, sensitive to the grease input to the extent that it tends to prevent oxygen diffusion through the cell membrane, disrupting the respiratory processes of microorganisms.

They are also sensitive to strong suspended solids, as these materials will rapidly tend to filter clogging. Physical pretreatment is very important, as it reduces the fouling of trickling filters.

When operating trickling filters, frequent and short-duration pauses are taken when the applied hydraulic loading rates are low. This allows a better adaptation to the daily variations of flow rates.

One of the problems raised by the operation of trickling filters is the proliferation of flies and mosquitoes in the hot season. Forced ventilation is not always sufficient, and applying intermittently a chlorinated solution to the filter may sometimes be necessary. The dosage must be calculated to avoid killing bacterial flora. Some plant managers recommend DDT powder against insects.

One of the advantages of trickling filters is the possibility to continue the operation even in the case of severe frost.

When an industrial effluent is connected to the treatment plant, the nature of the products spilled in it should be taken into account. This involves food processing industries, slaughterhouses and several pharmaceutical and textile industries, for which the content in organic materials is significant and treatment is possible by correctly dimensioning beds.

As for industrial effluents with a high proportion of acidic or basic mineral materials, poor operation of the biological bed can be expected in all cases.

The optimal pH for the operation and development of microorganisms ranges between 6 and 8. It should not drop below 5.5 nor exceed 9. The biochemical effect of pH is explained by an effect on salt concentration, which increases following the pH or by an enzyme action.

3.2.6.1. *Applicable organic and hydraulic loading rates*

Table 3.3 presents several parameters for the dimensioning and implementation of organic and hydraulic loading rates depending on the type of support material.

Parameter	BOD_5 degradation	BOD_5 degradation and nitrification	Nitrification
Support material	Rock or plastic	Rock or plastic	Rock
Feed type	Primary settling tank	Primary settling tank	Clarifier
Hydraulic loading rate ($m^3.m^{-2}.d^{-1}$)	14.7–88	14.7–88	35–88
BOD_5 loading rates ($kg.m^{-3}.d^{-1}$)	0.32–0.96	0.058–0.24	–
$N\text{-}NH_3$ loading rates ($g.m^{-2}.d^{-1}$)	–	0.2–1.0	0.5–2.4
Output concentration ($mg.L^{-1}$)	23–30 in BOD_5	< 10 in BOD_5 < 3 in $N\text{-}NH_3$	0.53 in $N\text{-}NH_3$
Height of the material (m)	< 12	< 12	< 12

Table 3.3. *Parameter dimensioning for trickling filters*

3.2.6.2. *Aeration of trickling filters*

Air circulation inside trickling filters is important for proper operation of the treatment process. It may be natural or forced. Natural ventilation, which is most commonly used, is vertically directed in the filter following air density (upwards or downwards). According to an observation, a 6°C temperature difference between the external air and the air inside the filter leads to a (downwards) air circulation velocity equal to 0.3 m^3 d'air.m^{-2} surface.min^{-1} or a velocity of 18 $m.h^{-1}$. The upwards or downwards circulation phenomenon caused by the difference of temperature between external air and internal air can be influenced by various external factors, such as wind. Indeed, if a strong wind blows horizontally at the top of an open trickling filter, it may exert a pressure variation, creating an air vent from the bottom to the top or vice versa. In this case, the natural aeration due to a difference of temperature can either be increased or reduced.

It proves, however, that the preferential air current along the trickling filter is the one occurring simultaneously with aspersion. In practice, the quantity of oxygen required for the degradation of organic materials is often below that brought by natural aeration.

Forced ventilation is justified by the specific trickling filter dimensions, such as, for example, deep filters or work under high loading rates. It is also adopted when the trickling filter must be covered to protect it against flies or to avoid the propagation of bad odors. In cold climate, forced ventilation by hot air leads to maintaining a proper temperature inside the trickling filter and preventing surface frost formation.

Forced ventilation is conducted with ventilators placed at the top of the trickling filter. They are expected to provide an air circulation of at least 18 $m.h^{-1}$.

3.2.7. *Calculations of trickling filters*

Many models for dimensioning trickling filters take into consideration various factors such as flow conditions, organic and hydraulic loading rates and the nature and the thickness of the biomass present on the material, as well as its physico-chemical properties.

Assuming as a first approximation that BOD_5 removal is in direct relation to the available biomass and the time of contact, the following can be written:

$$Lf/Lo = e^{-kXt}$$

where:

– Lo and Lf are, respectively, the input and output BOD_5 concentrations ($mg.L^{-1}$);

– X is the biological mass present on the material;

– t is the residence time of the effluent inside the trickling filter;

– k is a constant depending on the nature of the material, that of the substrate, etc.

Let us assume that the average contact time is:

$$t = k' \, H \, S/Q$$

where H is the height of the media support inside of the trickling filter, Q is the feed flow rate and k′ is a constant depending on the nature of the material.

Indeed, the residence time depends on various parameters such as the specific surface of the trickling filter, the temperature and the hydraulic loading rate. This leads us to writing the following relation:

$$t = (k'' \, H \, As^{b})/Q^{n}$$

with b = 1/3 and n = 2/3.

This relation is established by assuming that the sludge mass X is proportional to the specific area. Eckenfelder (1963) suggests taking into consideration the biological decline of the substrate according to a homographic function:

$$\frac{Lf}{Lo} = \frac{1}{1 + k''HAs^{b}Q^{-n}}$$

The parameter k″ is corrected depending on temperature:

$$k'' = k''_{20} \, 1.035^{(\Theta - 20)}$$

The National Research Council (1946) proposes the following equation for the reduction of BOD_5 on a single trickling filter:

$$E1 = \frac{100}{1 + 0.4432\sqrt{W1/VF}}$$

with:

- E1 being the reduction of BOD_5 at 20°C;
- W1 being the BOD_5 load DBO_5 ($kg.d^{-1}$);
- V being the volume of the trickling filter (m^3);
- F being the recirculation factor.

F is calculated as follows:

$$F = \frac{1 + R}{(1 + \frac{R}{10})^2}$$

R is the (dimensionless) recirculation ratio varying between 0 and 2.

In the presence of two serial beds, BOD_5 reduction in the second tier is given by:

$$E2 = \frac{100}{1 + \frac{0.4432}{1 - E1}\sqrt{W2/VF}}$$

with:

– E2 being the reduction of BOD_5 at 20°C on the second tier;

– W2 being the BOD_5 load ($kg.d^{-1}$) on the second tier.

The water temperature influence on BOD_5 reduction is calculated by:

$$E_T = E_{20}\ 1.035^{(T-20)}$$

EXAMPLE.– Dimensioning of a trickling filter.

– Data:

- wastewater flow rate: 7,570 $m^3.d^{-1}$;
- raw wastewater BOD_5: 200 $mg.L^{-1}$;
- expected BOD_5 at the outlet: 25 $mg.L^{-1}$;
- two serial trickling filters;
- height of the material in each trickling filter: 1.83 m;
- recirculation ratio: 2/1;
- temperature: 20°C;
- E1 = E2.

– Calculation of the efficiency of BOD_5 reduction: (200 – 25)/200 = 0.875 or 87.5%:

E1 + E2(1-E1) = 0.875, or then 1 = E2 = 0.646

– Recirculation factor:

$$F = \frac{1 + 2}{(1 + \frac{2}{10})^2} = 2.08$$

– BOD_5 load applied to the first trickling filter: 0.2 x 7 570 = 1,514 kg $DBO_5.d^{-1}$.

The volume is given by replacing E1 and W1 in the equation:

$$E1 = \frac{100}{1 + 0.4432\ \sqrt{W1/VF}}$$

Or:

$$0.646 = \frac{100}{1 + 0.4432\ \sqrt{1{,}514/V(2.08)}}$$

We obtain: V = 476 m^3.

Knowing the height, the surface can be deduced: S = V/H = 426/1.83 = 260 m^2.

The diameter of the circular trickling filter is 18.2 m.

The BOD_5 load applied to the second tier is:

W2 = (1 − E1) W1 = (1 – 0.646)1,514 = 536 kg.d^{-1}

The volume of the second trickling filter is deduced:

$$E2 = \frac{100}{1 + \frac{0.4432}{1 - E1}\ \sqrt{W2/VF}}$$

$$0.646 = \frac{100}{1 + \frac{0.4432}{1 - 0.646}\ \sqrt{536/V(2.08)}}$$

We obtain V = 1,345 m^3.

Therefore, the surface is: S2 = 1,345 m^3/1.83 m = 735 m^2 and the diameter is 30.6 m.

Summary:

– BOD_5 load first tier: 1,514 (kg.d^{-1})/476 m^3 = 3.18 kg.m^{-3}.d^{-1};

– second tier: 536 (kg.d^{-1})/1,345 m^3 = 0.40 kg.m^{-3}.d^{-1};

– hydraulic loading rate first filter: (1 + 2) 7,570/(24 x 260 m^2) = 3.63 m^3.m^{-2}.h^{-1};

– hydraulic loading rate second filter: (1 + 2) 7,570/(24 x 735 m^2) = 1.29 m^3.m^{-2}.h^{-1};

– dimensioning of plastic trickling filters: there are several formulae and models proposed in the literature. That of Water Environment Control (2000) is used here:

$$Lf/Lo = e^{-kH/C_h^n}$$

where Lo and Lf are, respectively, the input and output BOD_5 concentrations ($mg.L^{-1}$); k is a coefficient related to the plastic material $(L/s)^{0.5}.m^{-2}$; H is the height of the material (m); Ch is the hydraulic loading rate at the output of the primary settling tank ($L.m^{-2}.s^{-1}$); and n is the constant specific to the plastic material (generally n = 0.5). k varies with temperature according to the relation: $k_T = k_{20}\ (1.035)^{(T-20)}$.

Given that BOD_5 removal is determined as a function of the hydraulic application rate, the essential effect of the organic loading rate should be taken into account. Assuming that the efficiency of BOD_5 removal is equal for the same organic loading rate, the value of k has to be adjusted depending on the depth and BOD_5 concentration of the influent. The k value is normalized at a specified depth and at a BOD_5 concentration of influent as follows:

$$k2 = k1\left(\frac{H1}{H2}\right)^{0.5}\left(\frac{L1}{L2}\right)^{0.5}$$

with:

– k_2 being the normalized value of k on a site specific to the given height and an input BOD_5. k_2 is proposed as equal to 0.210 $[(L/s)^{0.5}.m^{-2}]$ (by DOW Chemical) (WEF 2000);

– k_1 being the value of k for a height of 6.1 m and an ingoing BOD_5 of 150 $g.m^{-3}$;

– L_1 being 150 g $BOD_5.m^{-3}$;

– L_2 being the BOD_5 ingoing on the trickling filter of the treatment plant ($g.m^{-3}$);

– H_1 being the 6.1 m height of the plastic material;

– H_2 being the height of the material on the treatment plant (m).

EXAMPLE.–

– Two-tier trickling filter of height: 6.1 m, in transverse flux in a plastic material of 90 $m^2.m^{-3}$. The starting point is n = 0.5.

– Flow rate: 15,140 $m^3.d^{-1}$.

– BOD_5: 150 $g.m^{-3}$.

– SS: 65 $g.m^{-3}$.

– T °C: 14°C.

– Expected outgoing BOD_5: 20 $g.m^{-3}$.

– Determination of k:

$$k2 = k1\left(\frac{H1}{H2}\right)^{0.5}\left(\frac{L1}{L2}\right)^{0.5}$$

$$k2 = 0.210\left(\frac{6.1}{6.1}\right)^{0.5}\left(\frac{150}{125}\right)^{0.5} = 0.230$$

– Temperature correction: $k_{14} = 0.230\ (1.035)^{(14\text{-}20)} = 0.187\ (L/s)^{0.5}.m^{-2}$.

– Hydraulic loading rate: $Lf/Lo = e^{-kH/C_h^n}$.

– We obtain:

$$Ch = [kH/Ln\ (Lo/Lf)]^{1/n}$$

$$Ch = [0.187\ (6.1)/Ln\ (125/20)]^{1/0.5} = 0.387\ L.m^{-2}.s^{-1}$$

– Surface of the trickling filter: S = Q/Ch = 15,140 x 1,000/(0.3875 x 24 x 3,600) = 452.2 m^2.

– Volume of the trickling filter: S H = 452.2 x 6.1 = 2,758 m^3.

– Unit surface of the circular trickling filter 452.2/2 = 226.1 m^2.

– Unit diameter: $(226.1 \times 4/3.14)^{0.5}$ = 17 m.

3.2.8. *Performances*

Table 3.4 presents the BOD_5 reduction rate for four types of trickling filters.

Although trickling filter processes are generally reliable, there is always a risk of operational problems that may be attributed to an increase in biofilm growth, poor design, a modification of wastewater characteristics or simply equipment failure.

Type of trickling filter	% BOD_5 removal
Low-loading-rate filter	80–90
Moderate-loading-rate filter	50–70
High-loading-rate filter	65–85

Table 3.4. *BOD_5 removal for various types of trickling filters*

3.2.9. *Advantages and drawbacks*

Table 3.5 presents the advantages and drawbacks of trickling filters.

Advantages	– Simple biological process. – Efficient for treating high concentrations of organic materials. – Suitable for small- and average-sized communities. – Good performances for BOD and COD. – Efficient nitrification. – Low energy consumption. – Moderate level of competence and technical expertise required to manage and operate the system.
Drawbacks	– An additional treatment may be required to meet stricter effluent standards. – Possible accumulation of excess biomass that cannot preserve an aerobic condition and may degrade the trickling filter performances (maximal thickness of the biomass is controlled by the hydraulic flow rate, type of media, type of organic material, temperature and nature of biological growth). – Requires regular attention from the operator. – Incidents due to clogging are relatively frequent. – Requires low-loading rates depending on the support. – Flexibility and control are limited compared to activated sludge processes. – Problems caused by flies, insects and odors. – Problems caused by snails.

Table 3.5. *Advantages and drawbacks of trickling filter*

Despite their age, trickling filters are still interesting for many constructors. Indeed, while trickling filters remain an expensive investment, their advantage is they only need low maintenance. This allows a significant reduction of workforce at the treatment plant. For wastewater quality equal to that of activated sludge, managing a trickling filter is simpler, as there is no need to manage internal sludge stock.

3.3. Biofilters

The most commonly used biological treatment processes are activated sludge (or free culture) systems, where waters to be treated are put in contact with a purifying biomass present in the form of suspended flakes, deeply mixed with the influent.

The increasingly severe effluent standards required for the preservation of receiving media and water resources, and consequently imposed by regulations, currently lead to the design of treatment plants as production units where maintaining permanent quality, reliability and treatment continuity is essential.

Activated sludge processes allow reaching these new performances, but at the cost of significant oversizing compared to 20 years ago.

Although simple and efficient in many applications, this technique has its limits related to the maximal concentration in biomass, which can be maintained in the system, to oxygen transfer and to the clarification stage, whose reliability is subject to the random quality of sludge produced (potential risk of filamentous bulking). Among these new processes (excluding membranes), processes with fixed biomass and mainly biofilters are among the most interesting. A fixed culture system is a reactor in which biomass is attached to a solid material, on whose surface water to be treated percolates. The fixed culture phenomenon is found in nature where biomass is fixed on pebbles, stones or rocks.

The purifying potential of fixed biomass systems is much higher than that of free biomass systems. This is explained by the fact that the selection of species and their concentration in the biological reactor takes place naturally by adhesion to the support material. This process can be used to select an active biomass attached in a higher quantity than that developed in activated sludge in the same volume of the reactor.

However, this potential can be correctly used provided that:

– there is possible adhesion and an insufficient quantity of the expected biomass (this aspect depends on support);

– there are no limiting factors, and it is possible to provide the needed oxygen to this biomass. In particular, the O_2 should be correctly transferred at each point of the biological support.

In a water treatment plant that is not a fermenter, even though there is a support facilitating biological adhesion, there is always competition between the species fixed on the support and the free species that develop between the support elements.

As oxygen can only be brought through the liquid layer, where the free species can be found, it is appropriate to prevent the concentration of this population, so that fixed species are not suffocated. The implemented technology, particularly in terms of the air transfer device and excess biomass removal, consequently has an essential influence on the performances of fixed biomasses.

3.3.1. *Principle*

The biofiltration principle relies on the use of a granular filtering material colonized by a purifying biomass and through which the treated effluent flows.

The technical processes of biofiltration, developed by various constructors, can be identified depending on their operation mode. The water feeding system can be placed at the bottom or at the top of the biofilter; water flows are then upwards or downwards. The characteristics of the grains that are part of the filtering material (nature, density, shape, etc.) are also specific to each constructor. The wide variety of materials (expanded clay, polystyrene, etc.) can be classified into two categories depending on their density.

Biofilters consist of a filtering medium where microorganisms are fixed and where colonization occurs. These microorganisms are responsible for the oxidation and for the removal of pollutants present in the water.

The biofiltration process is composed of several stages in the removal of water contaminants. Contaminants are transported to the bed medium in the aqueous phase.

Bacteria are immobilized on a support that offers the best conditions to proliferate. Wide surface support materials have various shapes.

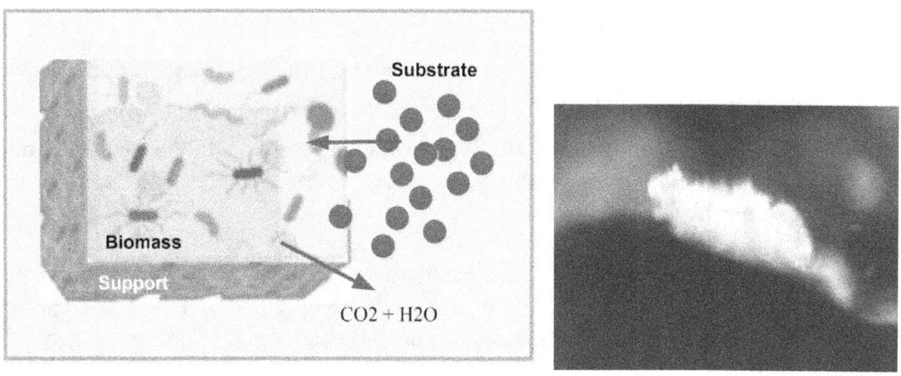

Figure 3.15. *Principle of degradation of the substrate on a biofilter*

The design of the biofiltration system is quite simple. A biofilter includes:

– a water distribution system where wastewaters are uniformly distributed on the surface of a biofilter according to an upwards or downwards circulation;

– a filtering medium;

– a floor supporting the material and allowing the passage of raw water (upwards current) or treated water (downwards current);

– aeration equipment.

In the case of a material whose density is above 1, the transit of the effluent to be treated through the gravel pack lying on a floor induces an expansion of the material (biofiltration in expansion, in the case of Biofor® and Biolest® up-flow biological filters).

In contrast, for this same material, the choice of a down-flow filtration involves high pressure on the floor, which bears the weight of the material, of the mass of water retained in the filter and of the biomass fixed on the support.

Conversely, the transit of the effluent through a material whose density is below 1, retained in the top part of the filter by a ceiling, induces the compression of the material (compression biofiltration, in the case of Biostyr® up-flow biological filters) (Gilles and Bourdon 1987).

The species responsible for the degradation of pollutants and which are fixed on the support depend on the operating conditions of the biofilter (oxidizer shape, type of pollution to be treated). Hence, the degradation of organic carbon facilitates the development of aerobic heterotrophic bacteria. In contrast, a limitation on organic carbon facilitates the development of an autotrophic biomass that can use a source of mineral carbon and ammonia nitrogen as a source of energy. These bacteria, known as nitrifying, oxidize the ammonia nitrogen into nitric nitrogen. Finally, under anoxic conditions, the presence of organic carbon facilitates the development of facultative aerobic heterotrophic bacteria able to use nitrates as oxidizer. These bacteria, known as denitrifiers, reduce nitrates into gaseous dinitrogen.

The biological treatment system may consist of three successive biofiltration stages expected to reduce carbon pollution, nitrification and denitrification of wastewaters. If the denitrification stage is at the end of the system, it will require the addition of significant quantities of carbon substrate, a carbon intake that may represent up to 40% of the operational cost of biofiltration units (case of the Seine Centre treatment plant). The consumption of carbon substrate upon downstream denitrification can be reduced by integration at the beginning of the biological treatment of a denitrification stage (Rocher et al. 2006, 2008).

In this case, a part of the nitrified waters is recirculated at the head of the biological treatment to be denitrified. The biodegradable organic material brought by settled waters meets the needs of the denitrifying biomass and, consequently, the operator no longer has to add any carbon substrate in this pre-denitrification stage.

a) *b)*

Figure 3.16. *Biofilters: a) open; b) covert*

This type of treatment is characterized by the fact that the biological reactor treats carbon and nitrogen but plays a filtration role, allowing the retention of excess sludge accumulated inside the reactor (generated biological sludge) and also of suspended materials carried by wastewaters. Hydraulic cleaning must be conducted periodically to remove excess sludge. These processes operate discontinuously and require several biofilters in parallel to ensure continuous water treatment; otherwise, a tank would be needed to store the flow rate during washing.

Biofilters for domestic water treatment can be used in anoxic (denitrification) processes or in aerobic (removal of carbon pollution, nitrification) processes. In aerobic operation, a transfer device is required for bringing the oxygen needed for biological reactions (Gilles 2004).

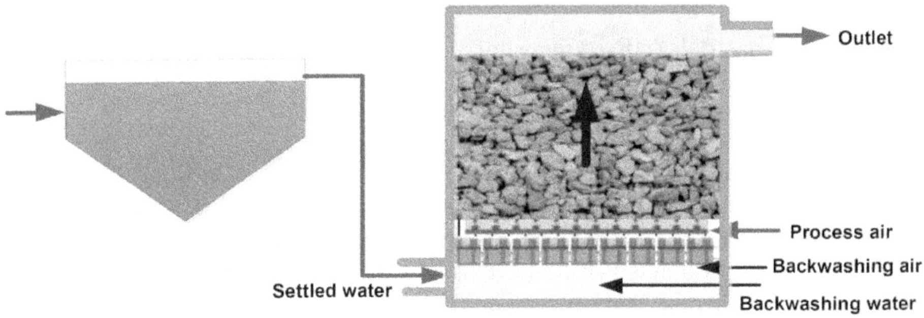

Figure 3.17. *Operating principle of a biofilter*

3.3.2. *Implementation and choice of material*

Biofiltration aims to simultaneously undertake the biological reaction of degradation of the pollution by the purifying biomass and the clarification by filtration of the treated effluent. It involves no final clarification by settling. The structure involves a filter having as media support a material whose granulometry is sufficiently small to obtain an efficient filtration effect, the material serving simultaneously as support to the fixed biomass.

The choice of material is a compromise between two opposite constraints:

– a sufficiently small granulometry to have a wide surface of adhesion and better SS retention;

– a sufficiently large granulometry to prevent too frequent clogging.

All proposed materials have common properties: they develop a large specific surface and have significant macroporosity. They have a high resistance to mechanical wear (to avoid problems of support attrition and of too strong a break of the biomass during washing). Some materials have a more or less significant adsorption capacity, which, by local concentration of the carbon pollution, accelerates the biological kinetics (Monod's law). This process is the most important with the use of carbons, and it is worth considering only for highly diluted waters.

There are various families of usable materials:

– more or less activated supports (carbons, alumina);

– calcined expanded natural materials (clays, schists);

– synthetic materials (polystyrene, polypropylene, etc.).

Ordered structures do not belong to this last category.

The shape of the material is also very important: in terms of attrition, a spherical material is more resistant than an angular material. It will also have a greater fraction of interstitial void space (free volume between the grains of material), which will allow operation at higher hydraulic velocity; conversely, its efficiency in retaining SS will be lower (Gilles 2004).

Expanded clays include:

– Biolite, Degrémont patent: small granules of expanded clay calcined at high temperature and presenting a significant open macroporosity.

– Expanded schist: small granules of expanded clay calcined at high temperature and presenting a significant open macroporosity.

– Biodagene, OTV (Veolia) patent: crushed slate that is dry-expanded in a furnace.

Synthetic materials include:

– expanded polystyrene beads, OTV material;

– plastic (polypropylene or polyethylene) beads.

Natural materials include Biodagene V, an OTV (Veolia) material consisting of black pouzzolane (particularly small expanded clay granules calcined at high temperature and presenting a significant open macroporosity, resistant to attrition) and naturally expanded material.

The material must have a high porosity and a specific surface (300–1,000 $m^2\ m^{-3}$) to facilitate the homogeneous distribution of the effluent, and mechanical and thermal stability to prevent the compaction of the filtering bed. Given the small granulometry of the material, a significant surface can be developed per unit volume, and consequently, it is possible to fix a large number of active microorganisms. The very small size of spaces between the grains coated with fixed active biomass allows the retention, by the filtration effect, of the suspended solids carried by the influent and the free biomass generated by the system. The nature of the material influences the fixation of species (Gilles 2004).

Figure 3.18. *Materials used in biofilters: a) pouzzolane; b) active carbon; c) expanded clay; d) biostyrene; e) schist; f) crushed brick; g) biozolane; and h) plastic support*

In such a system, the generated sludge retained in the material will gradually clog the biofilter. The material should therefore be cleaned cyclically. Unlike standard processes, this technique is discontinuous. As previously mentioned, several parallel filters ensure water treatment without interruption. The washing mode is generally issued from the techniques applied in sand filtration and adapted to the type of concerned filter. Waters used for washing filters are, most of the time, biofilter-treated waters.

Restarting the bed after correctly conducted washing is generally quasi-instantaneous. After washing, there is a sufficient adherent bacteria population to immediately obtain a treated water quality corresponding to the established regime. If correctly conducted, washing allows the removal of the retained interstitial sludge but should not have a too significant abrasive effect on the adhered biomass.

The support material can be in the form of bulk or of an ordered structure. Ordered structures are characterized by a high interstitial void space index and a low filtering capacity, due to which they are usable only in a refining treatment or completed by a unit that stops SS, the bulk material being mainly characterized by its granulometry, its nature and its density. The height of the material, depending on the type of biofilter or its application, varies between 1.5 and 5 m.

For all these materials, an effective size d_{10} is defined, representing the size of the opening through which 10% in weight of material passes. Proposed granulometries vary between 1 and 10 mm depending on the materials and on their use.

The diameter d_{50} is obtained by the relation: $d_{50} = ES.UC^{0.834}$.

The diameter d_{90} is given by: $d_{90} = d_{10}\ 101.67^{\log UC}$.

ES is the effective size of the media support.

The uniformity coefficient (UC) is the ratio between the diameter that lets 60% of particles pass and the one that lets 10% pass:

$$UC = \frac{d60}{d10}$$

where d_{60} is the size of the opening through which 60% of the material passes. This coefficient gives an indication referring to the granulometric homogeneity of the filtering mass. The values of the UC must be below 1.6 and are generally between 1.3 and 1.4. Lower values of the UC render the material more costly, as a high proportion of thin material is lost during washing, and higher values would reduce the porosity of the material.

Porosity ε is defined as the fraction of void volume compared to the total volume of the material. Therefore, it represents the proportion of void existing between the grains of stacked materials. This is why the volume occupied by the material is (1ε). Depending on the nature of the porous medium, ε varies between values near 0 to 1. The higher ε is, the lower the resistance to water flow. The porosity of a material does not depend on its size but only on its shape and its geometric characteristics.

Weight loss under fire at 450°C and weight loss when washing with hydrochloric acid (20% v/v over 24 hours at 20°C) must be < 2%.

The apparent density is the ratio of the material mass to the total volume occupied. It is therefore the mass of the dry bulky material occupying the unit volume.

In reality, the apparent density is a dimensionless parameter, as it is the ratio of density to the mass of the same volume of water (at a given T °C). It is equal to a thousandth of the density. The absolute density of the material, which is the volume of excluded voids, is far higher.

The density of the material must be sufficiently different from that of water to prevent losses during washing.

In terms of density, there are two groups of materials:

– dense materials (density > 1);

– floating materials (density < 1).

For dense materials:

– the higher the densities, the stronger the necessity of significant energy contributions to washing;

– the lowest densities generally correspond to a low mechanical resistance when facing abrasion, therefore an attrition risk and a higher risk of loss during washing.

For floating materials:

– high densities correspond to the most expensive materials: these floating supports are expanded synthetic materials; the denser they are, the higher the amount of raw material;

– low densities correspond to a low mechanical resistance with the risk of crushing of the material.

The size of grains conditions the specific surface of a material that is the sum of all external surfaces of the particles contained in a unit mass. The specific surface of a material As (m^{-1}) is the surface developed by a grain per volume of this grain.

For a sphere, it is equal to:

$$As = \frac{6}{d}$$

A grain sphericity factor measures the similarity of its shape to that of a sphere:

$$\psi = \frac{\text{volume of the grain}}{\text{volume of the circumscribed sphere}}$$

If the grain is nearly spherical, it fills nearly all the circumscribed sphere. On the contrary, if the grain looks more like a thin disk, it occupies a small piece inside the sphere. Sphericity increases as the grain occupies increasingly more space. The sphericity factor is 1 when the two volumes coincide. If a grain is not spherical, it has more surfaces for the same volume and consequently offers more resistance to transportation.

The hardness of the material is defined by its resistance to attrition in its application to filtration. The hardness measure on the Mohs scale depends on the capacity of a material to scratch a reference material or be scratched by a reference material. The measure is based on a scale from 0 to 10, the latter value corresponding to the diamond that can only be scratched by another diamond. Hardness should preferably be > 3 to make sure that washing operations do not produce fine particles.

Material	Dimension (mm)	Density (kg.m^{-3})	Uniformity coefficient	Specific surface (m^2.m^{-3})
Biodagene	3–6	< 1,300	1.4	800–1,200
Schist	3–5	1,300	1.3	800–1,000
Active carbon	3–6	800–1,000	1.3	> 1,200
Pouzzolane	3–6	1,200–1,300	1.4	900
Biozolanne	3–6	1,200	< 1.3	900–1,000
Biostyrene	3.6–5	45–50	< 1.2	1,000–1,200
Biolite	2.5–3.8	1,200	< 1.2	1,200

Table 3.6. *Characteristics of several materials used in biofilters*

3.3.3. *Biofiltration alternatives: operation in immersed fixed filters and fixed trickle filters*

The various biofilter alternatives differ by granulometry, nature of support, filtration direction, air transfer technology, operation type and washing mode.

A hydrostatic load is required to compensate for the head loss through the support material and allow filtration in almost all alternatives; this load is provided by an immersed bed operation. The structure of the biofilter must allow the preservation of the material in the system while allowing the passage of fluids (water and air) that cross it. A heavy material will be retained by a floor and a floating material by a ceiling. The proper operation of biofiltration requires an equal distribution of fluids in the biological reactor (to avoid dead zones). The filtering floor or ceiling is generally equipped with nozzles distributed in sufficient numbers on the filtering surface to allow this equal distribution and let fluids pass.

Equal-distribution parts are sensitive to clogging. This risk is minimized if they are located in a properly treated water environment (few biodegradable residual pollutions and few SS). Conversely, this risk is maximized if they are located in an untreated water environment (biomass development and SS accumulation).

For small filters, a more rustic structure consisting of a grating supporting a layer of pebbles (itself supporting the material) is used as a replacement of the floor with nozzle.

A design for the biofilter allowing an equal distribution of the hydraulic flux without clogging the distribution bodies is one of the key technological elements in biofiltration.

3.3.3.1. *Direction of filtration*

Filtration can be upwards or downwards. For aerated biofilters, with air always being up-flow, the first case will be that of air/water co-current flow and the second case counter-current flow.

In an installation with a filtering floor, upwards filtration leads to an operation allowing expansion of the material. Downwards filtration leads to an operation with compaction of material (Gilles 2004). The opposite occurs in an installation with a filtering floor where upwards filtration leads to compaction of material.

Material expansion operations allow the application of higher hydraulic flows but to the detriment of filtration performances (high leakage of SS in the treated water). Operations in the direction of material compaction allow the best filtration performances but limits hydraulic capacities.

For aerated filters, counter-current flow filtration can create a filter embolism (or hydraulic clogging): it is filter clogging when water speed is too high and air can no longer pass. This embolism phenomenon does not occur in co-current flow filtration.

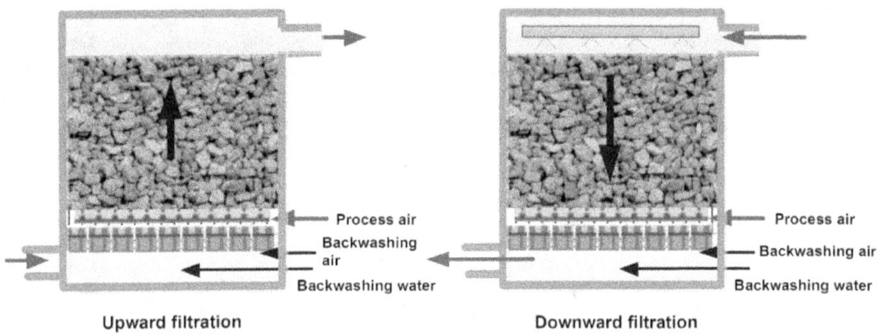

Figure 3.19. *Biofilter with two filtration directions*

3.3.3.2. *Aeration of biofilters*

In aerobic biofilters, biofilm development on the material requires a constant oxygen input. Many techniques are available on the market.

The required oxygen can be transferred by direct air draught in the biofiter, i.e. in the biological reactor. This mode ensures the best transfer efficiency, which is practically used by most constructors. Consequently, this is what will be found in most current biofilters.

Air draughts are generally undertaken with air ramps perforated with holes, sometimes coated with elastomer sleeves. To avoid dead zones and optimize biofiltration, the entire fixed biomass must be sufficiently fed with dissolved oxygen, which requires correctly conducted oxygen transfer at any point of the filtering mass.

The design of the oxygen transfer system is consequently another key technological element in biofiltration. Aeration can be undertaken either at the bottom or inside the bed. Potential combinations – filtration direction and air admission location – provide the following four operating possibilities (Figure 3.20).

3.3.3.2.1. Diagrams 1, 5 and 6: air–water co-current flow, aeration at the bottom of the filter

With a *dense material and filtering floor*, the operation is in the direction of decompaction of the material. This allows high water velocities but leads to maximal SS leakage. Moreover, the hydraulic flow-distribution parts are in an environment of untreated water and air. The risk of their clogging by a generated biomass cluster and SS is maximum.

With a *floating material and filtering ceiling*, the operation is in the direction of compaction of the material. Velocities that are applicable in water are reduced, but SS retention efficiency is increased. The hydraulic flow-distribution parts are in an environment of treated water. The clogging risk is minimum.

3.3.3.2.2. Diagram 2: air–water counter-current flow, aeration at the base of the filter

With a *dense material and filtering floor*, this hydraulic system operates poorly. Air draught must compensate the hydraulic head loss generated in the biofilter. The head loss downstream of the floor being lower, there will be a preferential air passage in the treated water circuit.

3.3.3.2.3. Diagram 3: air–water co-current flow, aeration in the bed

With a *dense material and filtering floor*, we can make a similar comment as for Diagram 1 except for the following nuance: the hydraulic flow distribution parts remain in an environment of untreated water but without air. There is still a clogging risk, but to a lesser degree than in Diagram 1. This diagram corresponds to the Biofor® (Degrémont) configuration. The air ramp is placed in it above the floor with nozzles. It corresponds also to the B2A®-Veolia configuration.

With a *floating material and filtering ceiling*, the positioning of the ramp in the material allows the protection in the filter of an anoxic zone in an environment of untreated water and the nitrification and denitrification (Biostyr®) in the same structure.

3.3.3.2.4. Diagram 4: air–water counter-current flow, aeration in the bed

With a *dense material and filtering floor*, this design can be used to avoid the air passage in the circuit of treated water, mentioned in Diagram 3, provided that the air draught is at 30–40 cm from the filtering floor (depending on the applied water velocities). The result is a non-aerated zone between the air ramp and the filtering floor.

This non-aerated zone cannot be used as a denitrifying anoxic zone, as it is located in an environment of treated water and is therefore low in degradable carbon pollution. It has a refining effect on SS.

Furthermore, the operation being in the direction of material compaction, this diagram is the one that allows reaching the highest filtration efficiency. Distribution bodies (nozzles) are in an environment of perfectly filtered treated water; clogging risk is minimum. Velocities applicable in water are strongly reduced because of the operation in the direction of material compaction, a phenomenon intensified by the air–water counter-current flow leading to filter embolism at high velocity in water (Gilles 2004).

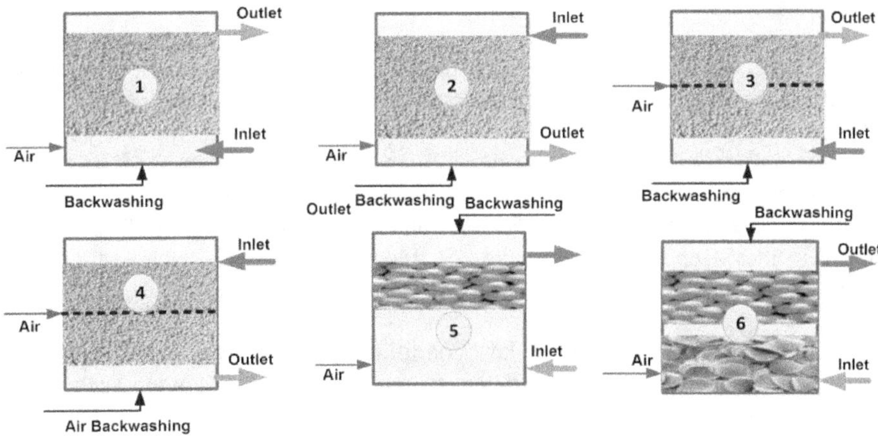

Figure 3.20. *Various configurations of biofilters with fixed and mobile supports*

3.3.3.3. *Backwash*

Because the filter gradually clogs because of the development of biomass and SS retention, filter backwashes are necessary. These backwashes can be conducted periodically (clock-controlled activation) or as set (maximum head loss).

The normal duration of a cycle (period between two backwashes) varies between 12 and 48 hours depending on the applied loading rate, granulometry and nature of the material. In most applications in domestic wastewater, one backwash per 24 h is scheduled, with this being clock-activated in the period of the daily hydraulic low (generally during the night), doubled for safety reasons by the possibility of a simplified wash activated as set.

Backwashing techniques resulted from those applied in sand filtration in drinking water; they use air and water. However, these techniques had to be adapted to the particular case of biofilters because of much deeper clogging than in the case of mechanical filtration, which is surface filtration.

Air serves to decompact the material and break the sludge and water to carry broken off materials. Depending on the alternative, we can use filtered water, settled flocculated water or filter feed water (when allowed by water quality), the latter case being very rare because of the risks of filter clogging. Depending on the type of process, washing will be conducted with co-current flow or counter-current flow for the filtration in the treatment phase. For biofilters designed with floor or ceiling nozzles, washing will always be conducted with sludge extraction opposite to distribution bodies, to avoid their clogging (Gilles 2004).

As counter-current flow washing has the advantage of directly removing retained SS without them crossing the entire filter, this washing design is preferred in Veolia techniques.

In the case of co-current flow washing and feed, the entire material will need to be well fluidized in order to remove SS, such an expansion requiring significant washing water velocities. Depending on the type of filter, one can use a backwashing water volume ranging between two and four times the volume of material.

A good backwash is another key condition for the proper operation of the biofilter. It must be sufficiently efficient to remove the "free" sludge accumulated between the grains of material; however, it must not be too efficient to avoid removing the biomass fixed on the material or leading to an abrasion of the material. An insufficient wash will lead to the accumulation of inert sludge in the system, to shortening cycles and to the smothering of fixed biomass. The final result will be diminished quality for the treatment and operational difficulties. Too vigorous washing will prevent the adherence of sufficient biomass to the material and will also lead to lower performances. In the case of attrition of the material, this will lead to a reduction of granulometry, whose long-term consequence is a shorter cycle duration. Optimum backwashing is specific to a given material (depending on its density and on its granulometry). It is also a function of the height of material and of the type of washing (co-current flow and counter-current flow). The durations for complete washing are in the range between 20 min and 1 h.

Depending on applied velocities, materials and granulometries, head losses after backwashing (clean filter) are in the range from 0.2 to 1 m and head losses before backwashing (clogged filter) in the range from 1 to 3 m.

After an efficient backwash, bacteria activity is instantaneously resumed. Approximately 30 min will be needed to reach the filtration potential, as the material has been decompacted and there is still a fraction (though small) of taken-off SS not removed during backwash which will be carried by the effluent at the beginning of the cycle.

Wash waters are dirty waters recovered once the filter is washed. These waters consist of untreated feed water in the zone above the biofilter and water used for washing. This phase drives the sludge accumulated in the biofilter. Foul waters are quite diluted (about 1 $g.L^{-1}$ of SS) and easy to thicken. They can be treated by a separation technique (settling or flotation) or sent back to the head of the treatment system.

3.3.4. *Dimensioning of biofiltration*

The various parameters encountered in the dimensioning of biofilters are the following (some are identical to those defined for activated sludge).

3.3.4.1. *Residence time*

Residence time represents the water passage time in the biofilter and is expressed in hours or minutes:

$$ts = \frac{V}{Q}$$

where V is the volume of the material and Q is the flow rate.

An average residence time is calculated over the daily average flow rate, and the peak residence time is over the peak flow rate. An apparent (average or peak) residence time takes into consideration V, the total volume of the biofilter (volume actually filled by the material + interstitial volume between the grains of material). In reality, water can only cross through the filter in the interstitial volume. The actual residence time involves the calculation of this parameter from the interstitial volume.

3.3.4.2. *Hydraulic loading rate or water velocity ($m.h^{-1}$)*

This is the ratio of the feed flow rate and the surface of the biofilter:

$$Ch = \frac{Q}{S}$$

where Q is the flow rate and S is the surface of the operating biofilters.

The terms used are average velocity, peak velocity or even velocity on N – n filters when n filters are washed.

3.3.4.3. *Volumic loading rate ($kg.m^{-3}.d^{-1}$)*

Cv represents a pollution flux per unit volume of material and per day:

$$Cv = \text{polluting flux } (kg.d^{-1})/\text{volume of material } (m^3)$$

The terms used are applied loading rate or removed loading rate; the volume of material might be the total volume or a specific (anoxic, aerated, etc.) volume. If a too significant load applied, it will be reflected by a lower efficiency and an increase in the concentration of the treated effluent. This loading rate is defined for each type of material.

In biofiltration, it is in fact impossible to quantify using simple methods the active mass attached to the material. It is possible to characterize and quantify what is extracted during backwashes but which, corresponding to the "free interstitial sludge", is not representative of the active mass.

If the filter is correctly backwashed, if almost all interstitial sludge is extracted from it, a backwash per day corresponds to a sludge age of about 1 day for interstitial sludge; this sludge age is not that of the active mass attached, for which growth, extraction and consequently sludge age are a natural process depending on backwashes.

3.3.4.4. *Air needs and velocity*

In aerated filters, the removal of carbon and/or nitrogen pollution consumes oxygen. This oxygen transferred in the liquid phase and then consumed will be most often provided, via transfer efficiency, from air draughts. The oxygen required to remove pollution is therefore reflected by flow rates of air blown in.

The terms used are daily air flow rate ($m^3.d^{-1}$), volume of blown-in air over the day and daily average flow rate if it is expressed in $m^3.h^{-1}$ (= Q daily air/24).

A peak air flow rate ($m^3.h^{-1}$) is also defined, necessary for dealing with the daily peak of pollution.

The O_2 needs (kg $O_2.d^{-1}$) concern the heterotrophic and autotrophic biomasses that can cohabit in the same biofilter. The O_2 needs depend on the quantities of biologically removed pollution.

With an organic material characterized by the COD and BOD parameters, the following approximations are used (Gilles 2004):

Required O_2 = a′ CO (or BOD) soluble influent + b′ $N.NH_4$ influent – c′ $N.NH_4$ influent – $N.NO_3$ effluent

The O_2 needs are defined with respect to the daily average values of pollutant fluxes. They are also defined in hourly values (kg $O_2.h^{-1}$) for the pollutant fluxes to be removed at peak hour. The oxygenation capacity to be implemented is established based on the peak needs.

The oxygen required for biological reactions is generally provided by forced drafts of air, via transfer efficiency:

$$O_2 \text{ to blow in (kg } O_2) = \frac{O_2 \text{ required x } 100}{\text{transfer efficiency (\%)}}$$

As 1 Nm^3 of air contains 0.3 kg O_2:

$$\text{Air to be blown in } (Nm^3) = \frac{O_2 \text{ to be blown in}}{0.3} = \frac{O_2 \text{ required x } 100}{0.3 \text{ x transfer efficiency (\%)}}$$

The following air velocities ($m.h^{-1}$) correspond to these air flow rates:

$$v \text{ air} = \frac{Qair}{S}$$

with Qair being the air flow rate and S being the surface of the operating biofilters.

As for the previously mentioned hydraulic velocities, the terms used are average aerial velocity and peak aerial velocity, peak velocities being defined with N or N – n filters when n filters are under wash.

O_2 consumed = COD removed – COD of bioformed sludge (hetero. + auto. + residue) + 4.57 N.NO_3 generated – 2.86 N. denitrified + O_2 dissolved effluent – O_2 dissolved influent

The material balance is conducted with a daily average value. The O_2 needs at peak times are deduced from it as a function of the hourly pollution peak coefficient admitted on the installation:

Peak pollution coef. = peak hydraulic coef. time sec x peak concentration coef. time sec

Peak O_2 coef. = Peak pollution coef. x 0.75 + 0.25

The peak pollution coefficient taken into account to define the peak O_2 coefficient is limited to a maximal value of 2, as biology cannot process more. This limits the peak O_2 coefficient to a maximal value of 1.75.

The air needs are deduced from the O_2 needs via transfer efficiency:

$$\text{Qair } (\text{Nm}^3/\text{h}) = \frac{\text{O2 (kg/h)}}{0.3 \text{ x transfer efficiency } (\%)}$$

The air needs are determined as a daily average value and hourly peak value.

3.3.4.5. *Sludge production*

The sludge produced is calculated as follows:

Sludge produced = sludge bioformed (hetero. + auto. + residue) + non-hydrolyzed influent SS

3.3.4.6. *Limits of dimensioning upon clogging*

Organic loading rates removed for the biofilter, applied hydraulic velocities, temperature and granulometry of the material determine the head losses required to ensure filtration and the head loss over time. The granulometry of the material determines a maximal applicable head loss.

The initial head loss (after washing), head loss and maximal applicable loading rate define the duration of the cycle.

The biological removal of pollution generates an increase in biomass, to which is added a particulate pollution mass retained by mechanical filtration. This accumulation gradually clogs the interstitial volume between the grains of material and increases the actual velocities in water and in air through the residual free interstitial volume.

The sludge produced (kg $SS.d^{-1}$) is the sum of non-hydrolyzed suspended materials brought by the (heterotrophic and autotrophic) suspended materials delivered by the influent and the (heterotrophic and autotrophic) biomasses produced by the biological process:

Sludge produced = influent non-hydrolyzed SS + generated heterotrophic biomass + generated autotrophic biomass

Excess sludge is the sludge recovered during washes. It corresponds to the produced sludge diminished by the SS leakage in the treated effluent:

Excess sludge ($kg.d^{-1}$) = sludge produced – effluent SS leakage

The generated heterotrophic biomass is a function of the quantity of biologically removed organic material and the autotrophic biomass of the quantity of nitrified nitrogen.

By identifying soluble pollutions and degradable and non-degradable mineral and organic particulate pollutions, the sludge mass becomes:

Sludge produced ($kg.d^{-1}$) = non-degradable VSS + non-hydrolyzed degradable VSS + non-hydrolyzed mineral SS + a removed BOD + b × nitrified nitrogen

with:

Non-hydrolyzed degradable SS = influent total VSS × degradable fraction of the particulate organic material × (1 – hydrolyzed fraction of degradable particulate organic material)

Hydrolyzed particulate organic material = total particulate organic material × degradable fraction of the particulate organic material × hydrolyzed fraction of the degradable particulate organic material

In a first approximation, it can be simplified by writing:

Sludge produced = k × influent SS + a removed soluble BOD + b × $N.NH_4$ removed

with $k < 1$ to take into account the hydrolysis of a fraction of particulate pollution.

This sludge is not stabilized and is very fermentable; the high content of volatile materials can be used for digestion or composting systems.

3.3.4.6.1. Head loss

A higher hydrostatic load is then necessary for filtration. The head loss between upstream and downstream of the filter increases during a cycle depending on the clogging degree. The head loss is expressed in centimeters of water column.

The initial head loss (HLi) is the head loss necessary after washing (clean filter free of its interstitial sludge with material seeded in fixed biomass) to allow filtration.

ΔHLt is the increase in head loss between the head loss at instant t and the initial head loss:

$$\Delta HLt = HLf - HLi$$

The final head loss (HLf) is the head loss required to allow filtration just before washing.

The head loss is mainly a function of:

– velocity in water;

– velocity in air (less strongly);

– height of the material H;

– granulometry of the material;

– void degree ε (free interstitial volume/total volume of the material);

– temperature.

The last point accounts for the fact that at high temperature, the biomass will develop from the first layers of the filter instead of developing throughout the bed, driving a more intense clogging localized on a limited height but responsible for a more significant head loss.

3.3.4.6.2. Duration of the cycle

It is the operating duration (in hours) between two backwashes. It is directly related to head losses. For a given material and a given design of biofilters, there is a maximal head loss (ΔHL max) that should not be exceeded:

– to avoid too strong clogging at the core of the system that would no longer allow obtaining efficient washes;

– to avoid the filter drilling: sludge leakage in the treated effluent;

– to avoid too strong pressure that would lead to the crushing of the material (this risk exists especially for filters with floating material operating in the direction of compaction of the material).

ΔHL max – HLi represents the maximal available head loss in the biofilter. It will allow the accumulation in the system of a certain quantity of sludge. This quantity of sludge is itself issued from the treatment of a given pollutant flux. The duration of the corresponding cycle is given by the ratio:

– treated pollutant flux leading to the available maximal ΔHL;

– admitted daily pollutant flux.

3.3.5. *Biofiltration performances*

The treatment quality results from three processes:

– a biological process of removing biodegradable pollution by the fixed biomass;

– a physical process of retention of particulate pollution admitted or generated in the biofilter, by mechanical filtration through the structure;

– a technological process.

3.3.5.1. *Biological process*

Similar to activated sludge, two types of biologies can be implemented and can cohabit in the same biofilter:

– The first one, via a heterotrophic biology, allows the removal of organic pollution.

– The second one, via an autotrophic (or nitrifying) biology, allows the removal of reduced nitrogen forms.

Only degradable pollution can enter the biological process. Biological reactions concern the degradable soluble organic material that titrates in COD and BOD and, according to the degrading mode, are divided into three categories.

The degradable soluble organic material includes the rapidly degradable soluble substrate, the slowly degradable soluble substrate and the adsorbable soluble substrate.

Biochemical kinetics develop as a function of the type of substrate:

– The rapidly degradable soluble one can be directly used by biology.

– The slowly degradable soluble one requires hydrolysis before being used. It remains in the soluble phase before degradation.

–The adsorbable one also requires hydrolysis before use, but it is adsorbed by biology. It is extracted from the soluble phase before degradation.

The degradable particulate organic material titrates in COD, BOD, VSS and SS. It requires hydrolysis before degradation.

The biological degradation of organic material requires nitrogen and phosphorus for the bacteria synthesis and also leads to a partial reduction of these parameters. The heterotrophic biology can also, under anoxic condition (absence of dissolved oxygen), remove the nitric nitrogen ($N.N0_3$). The nitric nitrogen then serves as a source of oxygen (replacement of dissolved oxygen).

The same is applicable for the soluble nitrogen material that titrates in NK and $N\text{-}NH_4$. $N\text{-}NH_4$ can be directly used by biology, and degradable soluble organic N requires a hydrolysis (or ammonification) before being used. It remains in a soluble phase before degradation.

Due to the short residential time of water (< 1 hour) and sludge (about 24 h) in the biofilter, hydrolysis kinetics are far more limiting than in activated sludge systems.

Overall, the biological process includes the removal of degradable pollutions, consumption of oxygen (dissolved or related to nitrates) and production of biomass that titrates in COD, BOD, SS, VSS, TKN and P.

There are several types of parameters that influence the biological performances.

3.3.5.2. *Physical parameters*

– Temperature: as for any biological process, biofiltration performances strongly depend on temperature. In practice, bacteria activity is multiplied by two when there is a 10°C temperature increase. At high temperature, biological kinetics is rapid, and the biological layer is concentrated at the beginning of the filter. The layers of material attacking the biofilter, strongly coated with biomass, are highly efficient in retaining SS. The leakage of SS to the filter bottom and in the treated effluent is low. At low temperature, the biological layer spreads deeper in the biofilter. The filtering capacity is lower, further driving SS to the bottom of the filter and increasing the leakage in the effluent.

– The dilution of the influent admitted to treatment: the concentration in the substrate plays a significant role in the biochemical kinetics of degradation. The higher the concentration, the faster the kinetics (Monod law).

– The variability of admitted pollutant fluxes: a well-fixed biofilm limits the leakage of microorganisms in the effluent. The biofilter is, however, more sensitive than activated sludge to short-duration variations because it is a reactor with a very short residence time and therefore low buffer capacity, and consequently, it cannot smooth the pollutant fluxes compared to a tank of activated sludge that has a long residence time. Biofiltration is therefore more sensitive to variations during the day, which is reflected by the lower quality of the peak treatment. The higher the variability during a day, the more significant is the deterioration of the effluent quality (at peak hours and as a daily average value).

– The support material influences the biological performances through its bacteria adhesion and solid biofilm formation capacity and its granulometry, which determines the available adhesion surface. By assimilating the material to spheres of a diameter equal to the average size of the material, the surface/volume ratio, in $m^2.m^{-3}$, is:

$$\frac{4\pi\left(\frac{d}{2}\right)^3}{\frac{4}{3}\pi\left(\frac{d}{2}\right)^3} = 6/d$$

By considering a void ratio ε between the grains, we have:

$$\frac{\text{surface of the media support}}{\text{apparent volume of the media support}} = \frac{6}{d}\ (1-\varepsilon) \text{ in } m^2.m^{-3}$$

The available adhesion surface is proportional to 1/d. The lower the average size of the media support, the larger the surface available for adhesion. The higher the interstitial void ratio, the more suspended materials cross the structure, reducing the efficiency of filtration. Consequently, the media supports ordered at a high void ratio have a very low efficiency. With high granulometry, the material has a reduced filtering capacity and leads to more significant SS leakage. The most commonly used granulometries in wastewater treatment range between 2 and 6 mm. With these sizes of media support, the purely physical filtering capacity is low. This is highlighted the first day of the start-up of a new biofilter, with efficiencies for SS that are generally below 50%. An efficient purely physical filtering capacity is obtained with lower granulometries; this is the case in the filtration of drinking water with granulometries of about 0.5 to 1.5 mm. In biofiltration, the filtering capacity becomes efficient when the filter is seeded. Therefore, filtration is not purely physical, being conducted through the biological layer fixed on the material. Filtration performances depend on the ratio between the flux of the SS to be retained by mechanical filtration and the generated biological mass that is fixed on the biofilter material. The lower this ratio, the higher the filtering capacity.

– Competition between biomasses: this applies especially to species with lower growth rate that would struggle to exist under conditions of competition with species at higher growth rate. It is the case with nitrifying autotrophic bacteria competing with heterotrophic bacteria removing carbon pollution. In tertiary nitrifying treatment, when there is little or no organic pollution to degrade, and therefore little generated heterotrophic biology, the entire adhesive surface of the material can be used by nitrifying species. This is not the same for nitrifying secondary treatment, where the two species of biomass must cohabit and share the available adhesion surface. In this case, the species with the highest growth rate will predominantly colonize the biofilter and will only leave a small space for the development of slower species that will have their growth limited. A balanced volume load between heterotrophic and autotrophic species will prevent a drop in the activity of bacteria that remove ammonia nitrogen.

3.3.5.3. *Technological parameters*

The proper running of a biological treatment with the help of fixed biomass depends on several considerations and technological parameters:

– Backwashing conditions: insufficient washing prevents taking off the excess biomass from the surface and may lead on a medium term to clogging. Too vigorous washing leads to detaching the entire biomass and reduces the substrate removal performances. Moreover, poor washing leads to the restart of the filtration phase, driving in the effluent SS that were not removed by washing, strongly deteriorating the filtration performances in instantaneous values as well as in the daily average.

– Aeration: this must be sufficient to transfer the oxygen in the environment and ensure the biomass needs. In the opposite case, aeration becomes a limiting parameter of the process, leading to a drop in biofilter performances.

– Equal distribution of the hydraulic flux: this conditions a homogeneous feed in the volume of material. The substrate being dissolved into wastewater, its availability for the entire bacteria mass is important. Poor distribution of the effluent leads to the formation of dry zones in which a part of the biofilm will lose its activity and therefore reduce performances. Poor distribution of the hydraulic flux leads to preferential circuits in the biofilter; in these preferential circuits, the velocity in water is increased, leading to a deterioration of filtration performances:

- too high hydraulic loading rates may lead to a leaching of the biofilter;

- high velocities in air increase turbulences in interstitial voids, decompact the material and also produce more significant leakage in the effluent;

- a significant height limits driving suspended materials to the bottom of the filter and reduces the SS leakage.

3.3.6. *Biofilter fields of use*

Within the same structure, biofilters can be used to biologically remove biodegradable soluble pollutions and retain, by filtration, suspended materials. The performances of one or another alternative can be very different, particularly as a function of specific efficiency in mechanical filtration and of the oxygen transfer capacity, specific to each technology.

The main functions of biofilters concern:

– removal of conventional pollutions (BOD, COD, SS) in secondary treatments as a replacement of classical biological treatment;

– simultaneous removal in secondary treatments of conventional pollution and nitrification or nitrification–denitrification;

– tertiary nitrification after an existing classical biological treatment;

– post-denitrification downstream of a nitrifying stage.

Concerning dissolved pollution, it is possible to remove pollution coming from carbon and nitrogen either simultaneously or successively.

Oxidizing reduced forms of nitrogen (nitrification) can be obtained simultaneously with the removal of carbon pollution:

– by strengthening aeration in the aerated tier of the biofilter;

– by reducing the carbon load admitted on the reactor.

Nitrification–denitrification can be achieved with two configurations: either by placing upstream of the aerobic tier an anoxic tier receiving all or a part of the primary settled water with recycling of the treated water bringing nitrates on the non-aerated tier or by achieving nitrification followed by denitrification with the addition of an external carbon source.

Where an additional phosphorus removal level must be guaranteed, in all cases this will involve the physico-chemical removal of phosphorus in pre- or post-precipitation.

Co-precipitation should be avoided (except for refinement) as it generates overproduction of physico-chemical sludge leading to too significant clogging.

Irrespective of the configuration of the filter, it will always be preceded by a stage that limits the input SS content in order to prevent too significant clogging.

3.3.7. *The place of biofilters in the wastewater plants*

Configurations to be considered depending on the objectives in terms of expected output quality of the effluent, treatment refinement and rehabilitation of an existing plant are summarized in the (not exhaustive) list below.

3.3.7.1. *Biofilter in secondary treatment*

– Primary stage:

- standard or lamellar settling;
- flotation.

– Secondary stage:

- carbon removal;
- nitrification;
- nitrification/denitrification.

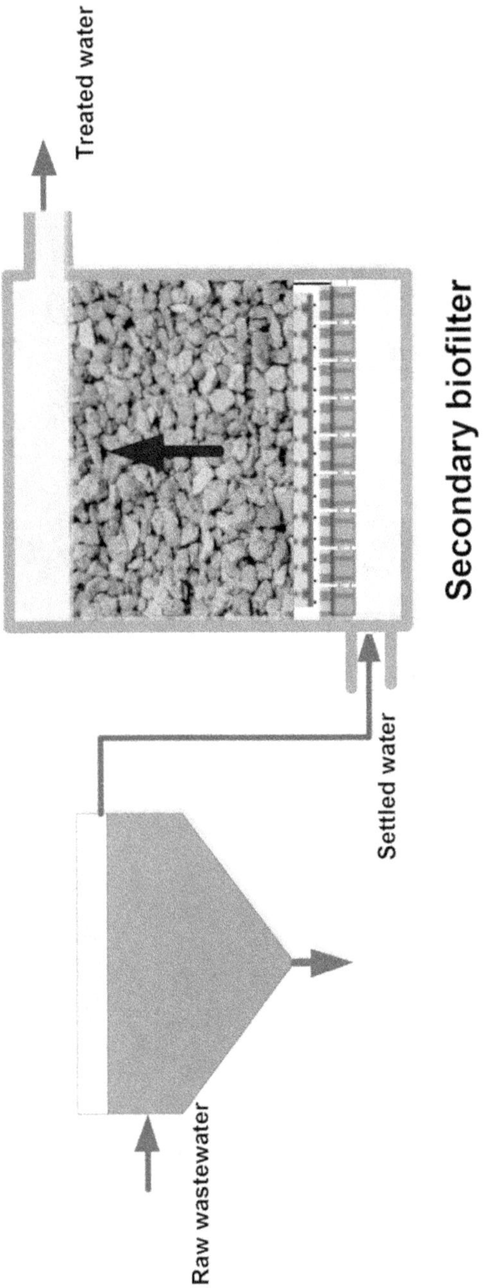

Figure 3.21. *Secondary biofilter in wastewater treatment plant*

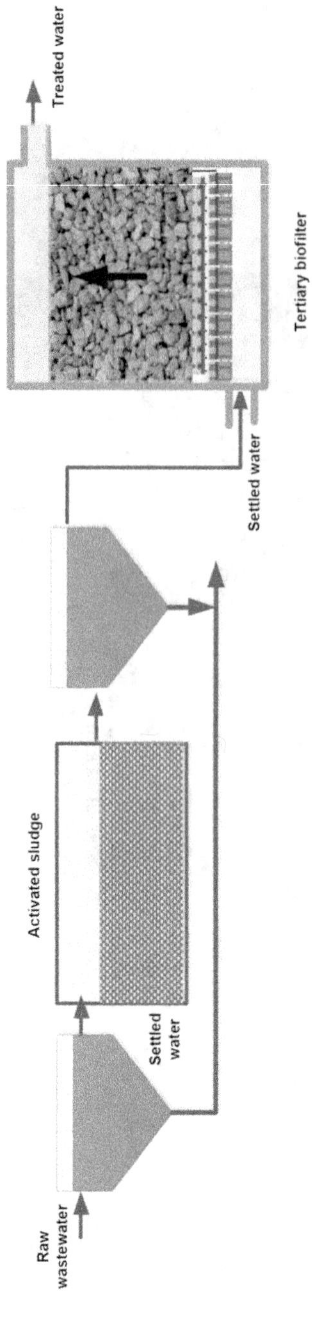

Figure 3.22. *Tertiary biofilter in the wastewater treatment plant*

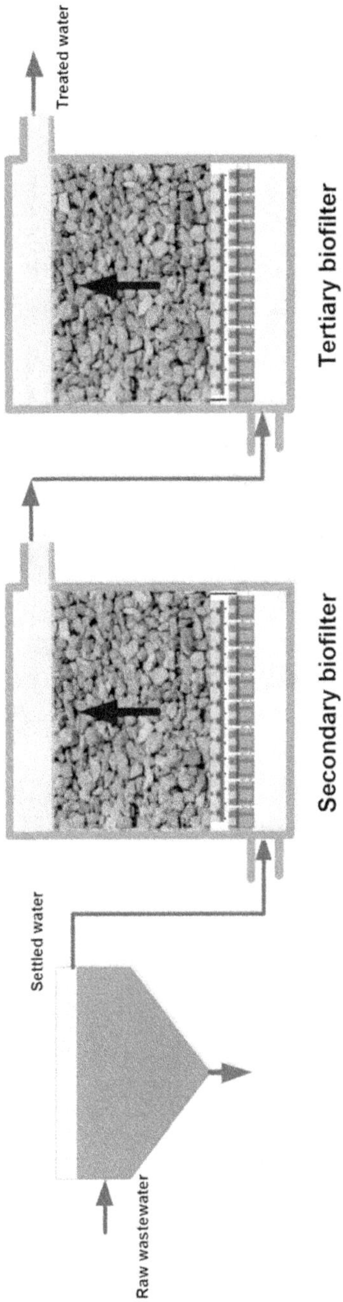

Figure 3.23. *Secondary and tertiary biofilters in the wastewater treatment plant*

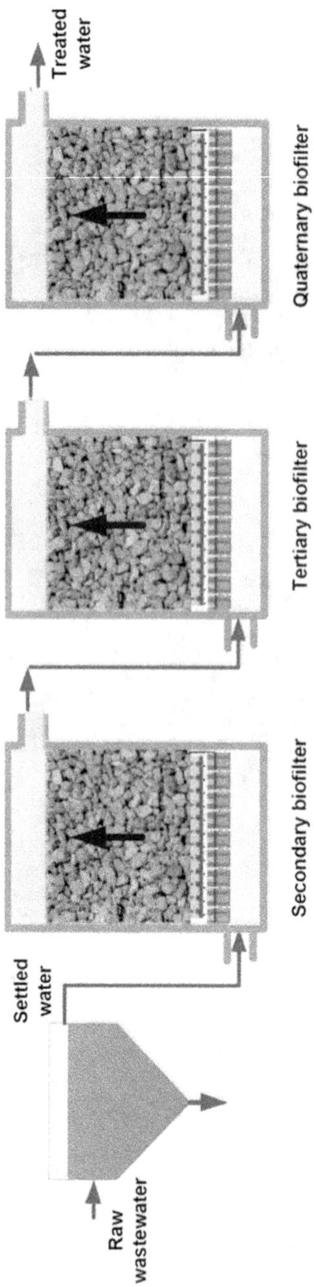

Figure 3.24. *Secondary, tertiary and quaternary biofilters in the wastewater treatment plant*

3.3.7.2. *Biofilter in tertiary treatment*

Tertiary stage:

– nitrification;

– nitrification/denitrification;

– denitrification with the addition of external carbon.

3.3.7.3. *Biofilter in secondary and tertiary treatments*

– Primary stage:
  - classical or lamellar settling;
  - flotation.
– Secondary stage:
  - carbon removal;
  - nitrification;
  - nitrification/denitrification.
– Tertiary stage:
  - nitrification;
  - denitrification with the addition of external carbon.

3.3.7.4. *Biofilter in quaternary treatment*

– Primary stage:
  - classical or lamellar settling;
  - flotation.
– Secondary stage: carbon removal.
– Tertiary stage: nitrification.
– Quaternary stage: denitrification with addition of external carbon.

3.3.8. *Arguments related to biofilters*

Compared to free culture systems, the advantages and drawbacks of biofilters are the following:

– Advantages:

- High treatment potential; they allow the application of very significant loads per unit volume and therefore address problems of surface coverage, structure coverage, etc. The volume of the reactor might be 4 to 10 times smaller than with activated sludge.

- In rapid start-up, fixation limits the leakage of purifying microorganisms in the effluent upon the start-up and, consequently, the optimum level of population in the reactor is very rapidly reached. This advantage allows addressing seasonal load variations.

- Safety for maintaining biomass in the reactor, to the extent that all the problems related to biomass settling are overcome.

- Operational stability due to the previous data when facing shock (load, temperature, toxic substances) effects.

- Quality of the treated effluent due to the conjugation of the biological effect and of an effect of biofiltration through the biological layer fixed on the support.

- System modularity that allows addressing the treatment reliability problems (splitting of structures), extension, variable capacity, etc.

– Drawbacks:

- The contact times being short, soluble pollution residues will be more than in extended aeration, and there will be little hydrolysis of the particulate pollution. This latter factor will especially influence denitrification, where it will often be necessary to add an external carbon source. On the contrary, activated sludge, in which particulate hydrolysis is quasi-total under low load, allows high denitrification performances without the addition of external carbon.

- The sludge produced being "fresh", the organic pollution content of the SS leakage will be more significant. It has a very fermentable character, risking a problem of odors on the sludge treatment row. In case of a problem, generated odors are predominantly mercaptan. The operator should very strictly manage sludge, avoiding extended residence in thickening or in storage, but this will not prevent odor emission in the dehydration room.

- While in extended aeration, the SS titrate approximately 1 kg of COD/kg of SS; in biofiltration, this value will rather be approximately 1.5.

- Because of short contact times, there is no buffer effect of the reactor; fixed culture systems will be more sensitive to fluctuations of the influent during the day. The consequence is wider variability of the treated effluent as well as the necessity to control the aeration in a more refined manner than in activated sludge.

- These processes require significant investment because, despite the gain of place, these processes require advanced automation, highly specialized feed and control systems and also specific materials.

3.3.9. *Performances of biofilters*

Biofilter performances can slightly vary depending on the type of material and on its physico-chemical characteristics. The geometry of the material influences the structure of the biofilm present on the support surface and, consequently, on the biomass-substrate surface. The diffusion and transportation mechanisms for the substrates to the biomass are the main limiting factors for the bacteria activity. The seeding of filtering materials is also widely influenced by the temperature of the environment; this parameter acts on the cell synthesis and also on the catabolism.

Given the fine granulometry and the capacity of materials to trap particulate pollutants, the clogging of biofilters can be rapid, leading to short cycle durations and imposing frequent backwashes. Primary physico-chemical settling and especially the dosage of reagents facilitate coagulation (ferric chloride, for example) and limit the flux of SS introduced. Lamellar settling thus efficiently removes particulate organic pollution. The SS contents in settled waters are therefore low (≤ 30 mg.L^{-1}), and the oxidizable organic material is mainly in soluble form: total COD, soluble COD and total BOD_5 are, respectively, 110, 70 and 50 mg O_2.L^{-1} (Rocher 2006).

Besides the retention of a large part of the SS, the effluent passage through the material is accompanied by an efficient removal of the biodegradable organic material. Hence, the residual organic pollution is treated upon the passage of settled waters through the biofiltration units. Irrespective of the used configuration, most of the carbon pollution is removed at the first stage. This stage ensures the physical filtration of particles (retention of 70% of SS – removed load of 0.5–1 kg MES.m^{-3} of material.d^{-1}). SS concentrations at the outlet of the first stage are low (< 10 mg.L^{-1}) and the next biofiltration units play only a minor role in the filtration of particles. Similar to SS, the biodegradable organic material is consumed at the first stage.

At the Seine Centre treatment plant (Rocher 2008), 85% of the BOD_5 admitted to the biofilters is consumed by the biomass of the first stage (removed load = 1.5–2 kg DBO_5.m^{-3}.d^{-1}) and the COD/BOD_5 ratio, an indicator of water biodegradability, is already of the order of 4 at the outlet of the first stage of biological treatment. The residual organic pollution not removed during the first stage of treatment is removed during the passage of the effluent through the nitrification units (removed load = 0.1 kg DBO_5. $m^{-3}.d^{-1}$) (Rocher 2006).

Nitrification efficiencies can be above 90% for NH_4^+ loads up to 1.2 kg $N.m^{-3}.d^{-1}$. This performance is explained by the low CODs/N-NH_4^+ ratio that feeds the nitrifying biofilter. This ratio of about 1.1 facilitates the nitrification kinetics and leads to high purifying efficiencies. Due to the application of a significant ammonia load, the nitrifying capacities of the filtering material can be maximally used. The nitric nitrogen formed during the oxidizing of the ammonia pollution is then reduced to atmospheric nitrogen. This denitrification stage can be integrated upstream or downstream of the biological treatment. In downstream denitrification, purifying performances are correlated with the added quantities of carbon substrate.

3.3.10. *Biofilter processes*

Name	Manufacturer	Filtration direction	Material	Type
Biocarbone®	OTV-Veolia	Down-flow	Biodagene Active carbon	Fixed
Biostyr®	OTV-Veolia	Up-flow	Biostyrene	Floating
B2A®	OTV-Veolia	Up-flow	Pebbles Gravels Biodagene	Fixed
Biostyr-Duo®	OTV-Veolia	Up-flow	Biostyrene Polyethylene chips	Floating and mobile
Biofor®	Degrémont-Suez	Up-flow	Biolite	Fixed
Biodrof®	Degrémont-Suez	Down-flow	Biolite	Fixed
Biolest®	Stereau-Bouygues	Up-flow	Biozolanne	Fixed
Biopur®	AFS Aqua	Up-flow Down-flow	Expanded clay	Fixed

Table 3.7. *Main suppliers of biofilters*

There are many biofilter processes proposed by the manufacturers. They are identified by the nature and granulometry of the support, air transfer technology, filtration direction, washing mode, applied loads and operational mode. Table 3.7 presents the main biofilters used.

Biofilters are divided into two categories:

– processes with a fixed bed;

– processes with a floating and/or mobile bed.

3.3.10.1. *Processes with a fixed bed*

The first biofilters designed implemented natural materials (clay type, expanded schist, etc.) of densities comparable to water ranging between 1.5 and 2. This is why the granular bed (1–4 m height) rests on a floor with nozzles, and the supply of water to be treated is either down-flow (Biocarbone® process, Figure 3.25) or up-flow (Biofor® process, Figure 3.35). There are many other types of biofilters such as Biobead® from Brightwater, Biopur® from Wabag, BAF® and Biolest® from Stereau. The granulometry of the material ranges between 2 and 6 mm depending on the expected treatment level.

To eliminate carbon pollution, the applied volume loads range between 2.5 and 3.5 kg BOD_5 per/m^3 of material per day.

Nitrification alone is conducted for lower BOD_5 loads (1.5–2.5 kg $BOD_5.m^{-3}\cdot d^{-1}$) and nitrogen loads ranging between 0.4 and 0.6 kg $NH_4.m^{-3}\cdot d^{-1}$.

It is possible to conduct nitrification–denitrification by using an anoxic filter that ensures the denitrification phase.

3.3.10.1.1. Biocarbone® process (OTV-Veolia)

The Biocarbone® process is a biological treatment process of the “fixed biomass” type that acts as a biological filter on dense material. Given the down-flow operation, it allows the biological degradation of pollution and clarification (biomass/water separation) by filtration (Figure 3.25).

Biocarbone® is a treatment process using bacteria fixed on a granular material aerated in the mass, which ensures both the bacterial degradation of pollutants and the retention of suspended materials. Its strong purifying capacities for removing carbon and nitrogen pollution have led to its adoption for over 150 projects throughout the world.

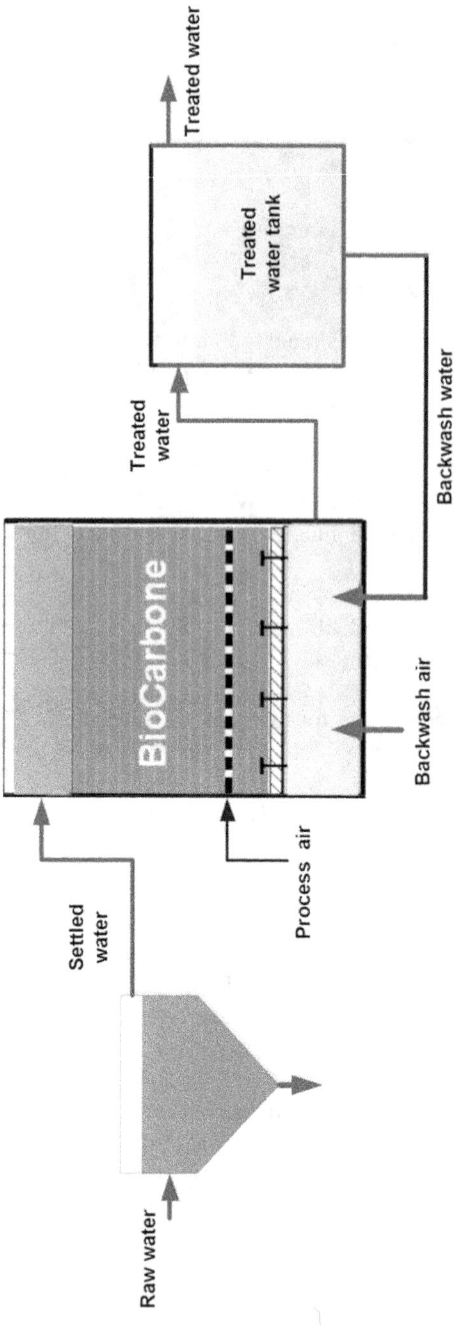

Figure 3.25. *Schematic diagram of the Biocarbone® (OTV-Veolia) process*

Technological elements

The filtering floor consists of small tiles that lead to unit surfaces of biofilters from 6.6 to 84.1 m^2. The floor is perforated and has nozzles with a density of 49 units/m^2. A layer of 8–12 mm of rolled gravel of 10 cm thickness is implemented above the floor to facilitate water drainage and protect nozzles from abrasion by the material during washes.

An aeration device blows the air through a ramp with holes of small diameter positioned at about 30 cm above the floor.

In the case of treatment aimed only at removing carbon pollution or low nitrification ($N.NH_4 > 5$ $mg.L^{-1}$), the air ramp has an aeration density of 80 orifices/filtered m^2, with orifices of diameter 1.5 mm.

In the case of high nitrification treatment ($N.NH_4 < 5$ $mg.L^{-1}$), the ramp has an aeration density of 160 orifices/filtered m^2 with orifices of 1.2 mm diameter. Air flow rates vary between 5 $m^3.m^{-2}\,h^{-1}$ and 40 $m^3.m^{-2}\,h^{-1}$.

The material constitutes a support resistant to mechanical wear while ensuring its "biological" role. It can be either from active carbon or expanded clay (biodagene) in two granulometries depending on the treatment qualities to be reached (2–5 mm for nitrification and 3–6 mm for carbon removal and denitrification).

Biodagene is an expanded material inside the grain and glassy at the surface that has an excellent resistance to friction (attrition). It is a crushed schist (slate) that is dry-calcined in a rotary furnace at 1,200°C. Its density (< 1.6) is sufficiently low to conduct efficient washes (for easy expansion of the material). Its specific surface is 800–1,200 $m^2.m^{-3}$. The retained granulometry is 3–6 mm for the removal of carbon pollution and from 2 to 5 mm for tertiary treatments and use under a nitrifying regime.

The natural flow of fluxes through the material, i.e. an upwards air flow and a downwards water flow leading to a system with counter-current flow, can be chosen.

This choice allows:

– obtaining a minimal head loss when the material is crossed by flows;

– obtaining equal distribution of flows in the structure.

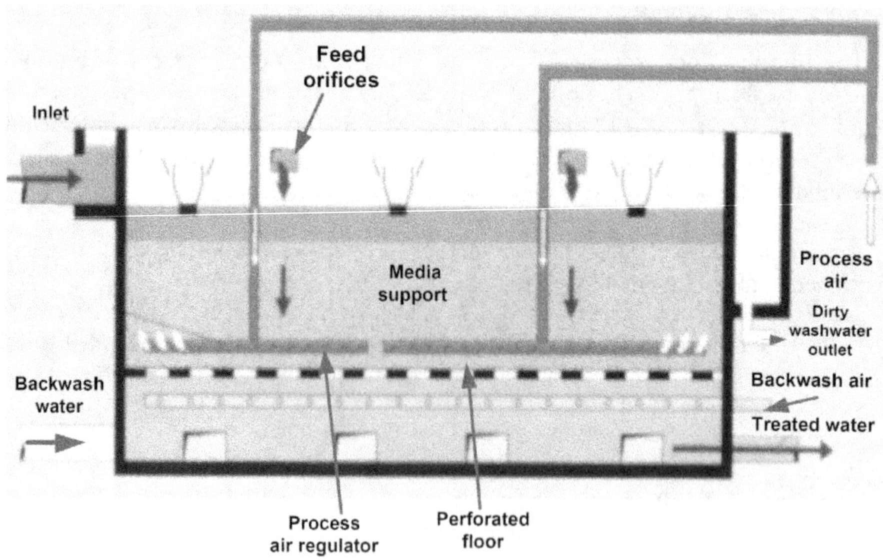

Figure 3.26. *Biocarbone® process equipment*

The height of the filter ranges between 1.8 m and 3 m. Air is blown into the material by a grid located at 30 cm above the filter basis. This device provides a good distribution of the air in the filter and, thanks to the pathway and to the breaking of the air bubbles on the grains of material, it is possible to have high oxygen transfer efficiency. It is the first biofilter where the transfer is made by blowing air directly into the biological reactor site, where the oxygen is consumed. This allows the best possible use of the blown oxygen. Moreover, filter aeration allows the penetration of materials to the core of the reactor and prevents surface clogging.

Aeration, which decompacts most of the media support of the filter, permits the penetration into the filter of suspended solids contained in the water to be treated and allows a significant storage capacity between two backwashes. The layer of material under the aeration grid has a refining role for the suspended solids.

To ensure water filtration in the process phase, a hydrostatic head loss is required between the water level above the filter and the outlet level. This head loss and clean filter depends essentially on the velocity applied in the water. It allows starting the backwash when needed. Level probes detect a threshold of admissible load and start the washing process fully automatically.

Indeed, biomass development and the retention of suspended materials produce gradual clogging of the biofilter and the necessity of periodic backwashes. These backwashes are conducted counter-current flow to the treatment/purifying operation and are composed of several successive cycles, each of which contains a decompacting phase for the material only by air blowing, a backwashing phase for the media support by air and water blowing and a final phase of water rinsing. The last phase allows the removal of dirty waters above the filtering material (by siphons or by special duct). Velocities in air and in water are 50 $m.h^{-1}$ in air and 50 to 60 $m.h^{-1}$ in water. Backwashing duration is about 20–30 min. The duration between two backwashes depends on the applied organic loading rates and the effluent temperature, and it varies from one to several days.

An efficient wash requires a water volume corresponding to about 2.5 times the volume of the backwashed material.

Backwash can also be started at regular intervals using clocks during pollutant load gaps or during the hours with lower energy costs. Automatic control through level probes is also possible and activates mini-washing if needed.

It is the water treated by the Biocarbone® filter that is used for backwashing. For this purpose, a treated water reserve of a minimum volume corresponding to water needs to backwash a cell is preserved in a tank containing purified water. Dirty waters are recovered in a second tank (dirty water tank) whose minimum volume corresponds also to the volume of water used to backwash a cell. These waters consist of extracted sludge, treated water (about 80%) and settled water (about 20%). These waters, after storage in a tank of dirty waters, are then sent to a unit of specific treatment or to the head of the station. After backwash, bacteria populations that remained fixed on the grains of material are sufficient to immediately obtain the quality of the treated water corresponding to the established regime.

Compared to classical activated sludge, the Biocarbone® process has the following advantages:

– single reactor: removal of the secondary clarifier, whose operation may be disrupted by the proliferation of filamentous species (bulking phenomenon) or, in the case of use in nitrification, by nitrogen gas emissions from uncontrolled denitrification;

– energy saving due to high oxygen transfer efficiency;

– upon rapid start-up, keeping the biomass in the aerated reactor of activated sludge processes is particularly difficult;

– very high quality of the effluent for the total suspended solid (TSS) criterion (of about 5 $mg.L^{-1}$), and therefore for the BOD/COD criteria associated with TSS, thanks to the biological filtration and to the filtering layer of non-aerated refining; this low content in suspended materials also allows a downstream disinfection with rates of chlorine or ozone 30% below those required for water treated by activated sludge;

– high purifying capacity, of about 3–4 kg of BOD per m^3 of material and per day (compared to less than 1 kg BOD_5 per m^3 of tank and per day for activated sludge, without taking into account the secondary clarifier), and high filtration velocity, of about 4 to 6 $m.h^{-1}$, which allows designing ultra-compact installations;

– sludge production is about 0.4 kg SS. kg^{-1} removed COD;

– the calculation of the blown-in O_2 input is based on an average of 10 kg O_2 brought per kg of COD to be treated. On site, the observed values range between 8 and 9 kg O_2 brought per kg of COD to be treated;

– power consumption for process air flow rates is about 140 $W.m^{-3}$ and 650 $W.m^{-3}$ for washing air;

– concerning aerated biological filtration techniques, the options retained for the Biocarbone® have the following advantages:

- filtration in the direction of compaction of the material and non-aerated specific filtration layer, which lead to excellent efficiency for the TSS criterion and the associated COD/BOD,

- direct supply of raw water after pretreatment (low content of organic substances in the wastewater) or primary settling, without micro-screening installation or equal distribution, with the device sensitive to clogging,

- nozzles only in contact with filtered water, eliminating any clogging risk;

– effective counter-flow washing. The filter zones with the highest load of suspended materials and biomass are downstream of the filter in the direction of air flows and washing water;

– immersion of the aeration duct in the filter, which allows an optimal equal distribution and better transfer of process air. As the aeration duct is in contact with the purified water, fouling risks are significantly reduced.

Performances

Carbon pollution

COD removal depends on the applied volume load (kg $COD.m^{-3}$ material. d^{-1}). According to Figure 3.27, beyond 7 kg $COD.m^{-3}$ material. d^{-1}, the output COD exceeds 90 $mg.L^{-1}$. An average load of 5.5–6 kg $COD.m^{-3}$ material. d^{-1} will therefore be needed. This load can be applied when the required standard for COD is 120 $mg.L^{-1}$. These results are obtained at a water temperature of 12°C. For a higher temperature, the standard COD is reached for loads exceeding 8 kg $COD.m^{-3}$ $material.d^{-1}$.

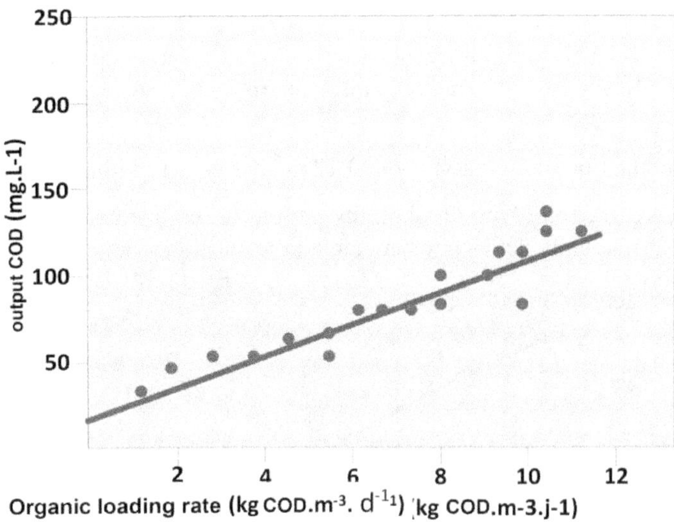

Figure 3.27. *Output COD as a function of the applied organic loading rate at 12°C*

Nitrogen pollution

The removal of nitrogen pollution involves several successive reactions:

– Ammonification, which is the transformation of organic nitrogen into ammonia nitrogen. This reaction unfolds to a large extent in the sewer system upstream of the treatment plant. There are various ways to remove the organic nitrogen residue in the plant: one part is trapped by settling in the primary treatment, another part by filtration in the Biocarbone® filter and the complement is transformed into ammonia nitrogen through the treatment plant, mainly at the level of primary settling. Generally speaking, organic nitrogen is quasi-totally removed downstream the plant.

– Assimilation: the synthesis of bacteria cells conducted during the removal of carbon pollution by biological treatment mobilizes nitrogen.

– Nitrification, which is the oxidation of ammonia nitrogen into nitric nitrogen (NO_3^-) in an aerobic environment. This reaction unfolds in two stages: *nitrification*, conducted by autotrophic Nitrosomonas bacteria that transform ammonia nitrogen into NO^- nitrate, and *nitration*, conducted by autotrophic *Nitrobacter* bacteria that transform nitrites into NO_3^- nitrate.

Three important characteristics of the nitrification reaction must be noted:

– Aerobic metabolism of nitrifying bacteria drives consumption of oxygen of about 4.25 mg of O_2 per mg of removed nitrogen: oxygen transfer must therefore be significant.

– Low production of biomass: about 0.17 mg per mg of oxidized nitrogen.

– An intake of about 7 mg of alkalinity per mg of oxidized ammoniacal nitrogen is required; an alkalinity deficiency may inhibit the reaction.

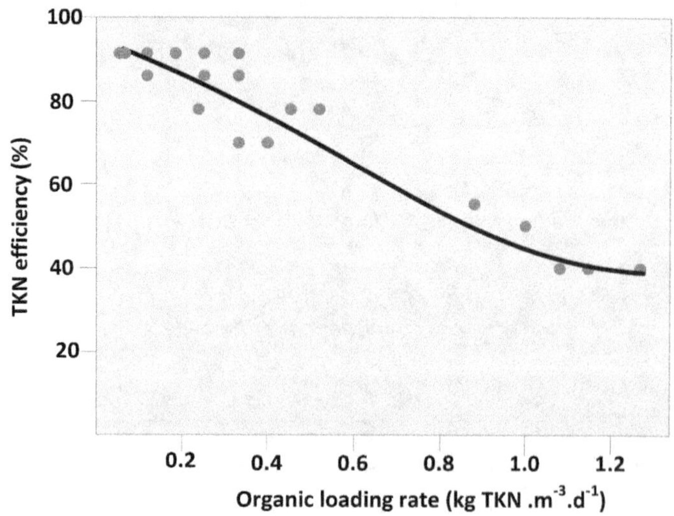

Figure 3.28. *Removal of TKN on Biocarbone® at 12°C*

COMMENT ON FIGURE 3.28.– *The increase in the load applied in TKN leads to a reduction of nitrification. The efficiency of removal for this parameter depends on many factors such as a good distribution of the air, water temperature, low carbon concentration and proper management of washes. Installations have shown that efficiencies above 80% in TKN are obtained for applied loads up to 0.45 kg NTK.m^{-3}. d^{-1} or 0.36 kg NH_4 per m^3 of material and per day.*

Because of the embolism problem, the hydraulic load is generally the parameter limiting the dimensioning. The hydraulic load limit depends on temperature. For the aerated filters, the hydraulic limit values of the process as a function of temperature are presented in Table 3.8.

Temperature	Limited hydraulic loading rates ($m.h^{-1}$)
8°C	7
18°C	5
20°C	4.5
25°C	3.5

Table 3.8. *Hydraulic loading rate limits of the Biocarbone® process as a function of temperature*

The analysis of velocities applied ($m.h^{-1}$) regarding the quality of the output effluent reveals a remarkable effect of biofiltration on the particulate, with efficiencies decreasing in the presence of high velocities. Indeed, the increase in velocity tends to decrease the passage time and thus limits the time of contact of the substrate to be degraded with the biomass.

For anoxic (denitrifying) filters, limited velocities are 15 $m.h^{-1}$ at low temperature and 12 $m.h^{-1}$ at high temperature.

	Carbon pollution	Secondary nitrification	Nitrification–denitrification	Tertiary nitrification
COD loading rate $kg.m^{-3}$total. d^{-1}	8	4	4	-
Nitrifiable loading rate $kg.m^{-3}$aerated. d^{-1}	–	0.4	0.6	0.65
Denitrifiable loading rate (upstream anoxic) $kg.m^{-3}$anoxic. d^{-1}	–	–	0.6	–
Denitrifiable loading rate with methanol (downstream anoxic) $kg.m^{-3}$anoxic. d^{-1}	–	–	1	–

Table 3.9. *Loading rates applicable to Biocarbone® at 12°C*

Provided that hydraulic constraints are met, the loading rates applicable at 12°C in domestic wastewater are presented in Table 3.9 to reach the following levels: COD = 80 $mg.L^{-1}$, BOD = 20 $mg.L^{-1}$ in carbon pollution, TKN = 5 $mg.L^{-1}$ in nitrification, $N.NO_3$ = 3 $mg.L^{-1}$ in denitrification output.

Fields of application

For domestic wastewater taken at any stage of a classical treatment system, the Biocarbone® process allows respecting all the regulatory standards concerning suspended materials, oxidizable materials and nitrogen materials.

The treatment after primary settling or a conventional biological process is applicable in the following domains:

– removal of only suspended solids and oxidizable organic substances (in secondary treatment);

– removal of suspended solids and oxidizable organic substances and nitrification (in secondary treatment);

– removal of suspended solids and oxidizable organic substances and nitrification-denitrification (in secondary treatment). In this case, a specific anoxic (non-aerated) structure is required for denitrification, with recycling of anoxic nitrates generated in the aerated structure (upstream denitrification);

– only nitrification (in tertiary treatment).

These treatments are applicable to any type of wastewater (urban or industrial) provided the influent is biologically degradable.

Phosphorus treatment is conducted on another structure of the water treatment system.

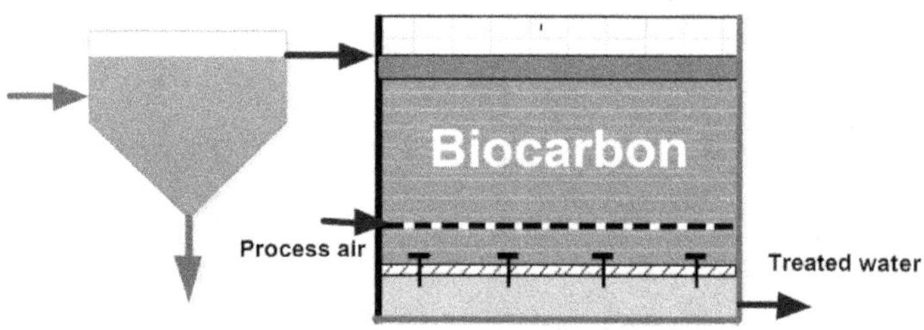

Figure 3.29. *Biocarbon with single carbon or carbon–nitrification process*

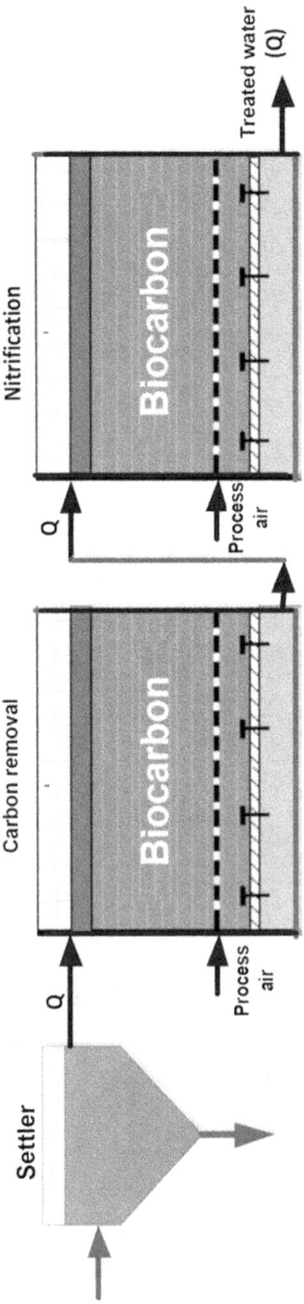

Figure 3.30. *Biocarbon with carbon process and nitrification in two stages*

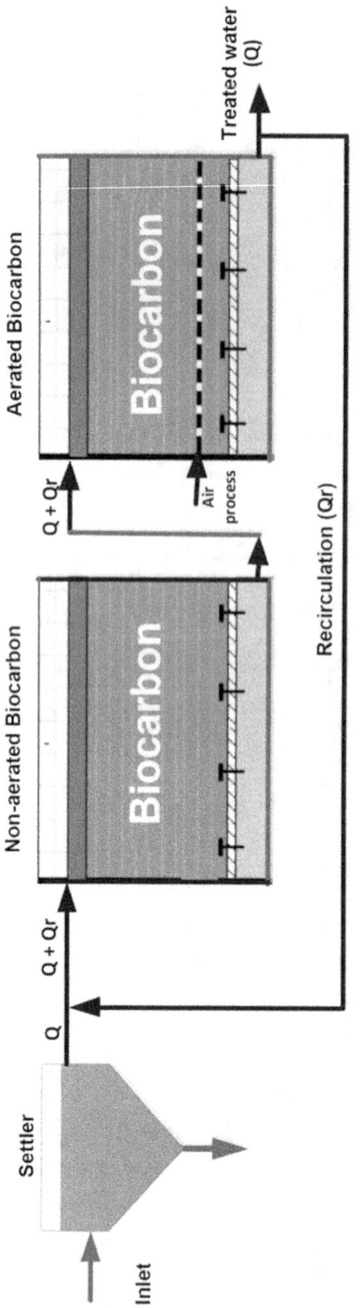

Figure 3.31. *Biocarbon in denitrification–carbon-nitrification configuration*

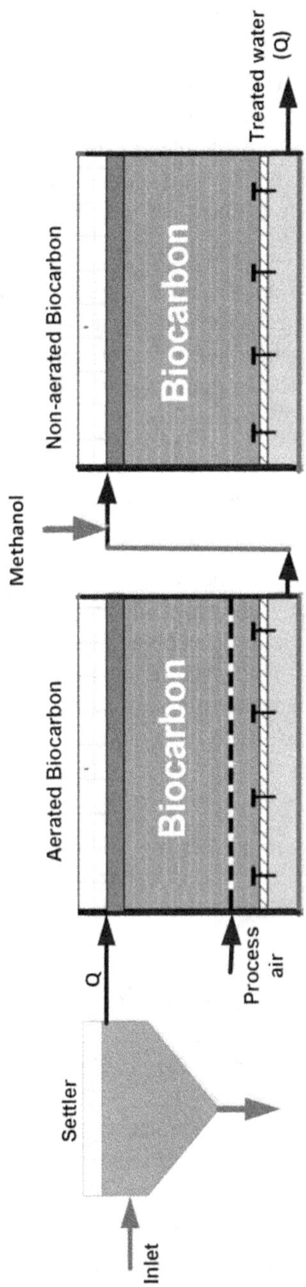

Figure 3.32. *Biocarbon in carbon-nitrification–denitrification configuration*

Specific advantages of biocarbon

– The SS quality of the treated effluent: it is the biofilter with the highest performance at this level, the residual SS ranging between 5 and 10 $mg.L^{-1}$. A particularly interesting visual aspect is the consequence, and it is possible to reuse the filtered water for industrial purposes: cooling circuits, declogging of filters, watering, etc. This low content in suspended solids facilitates efficient disinfection with a rate of chlorine or ozone about 30% lower than what is needed for wastewater treated by activated sludge. The clarity of the effluent also facilitates the disinfection by UV radiation.

– Optimal protection of parts sensitive to clogging: the nozzles of the filtering floor and the process air pipes are in an environment of treated water with a very low load of SS. Biocarbone® is the only biofilter that simultaneously has all this equipment in an environment of treated water. The clogging risk is minimum.

– Easy equal distribution of process air between several cells: the design of Biocarbone® drives an air pressure process in the distribution orifices of the air duct, independently of the degree of filter clogging. This facilitates the equal distribution of air between several operating cells.

– The protection of the material facing the risk of grease spreading: the layer of non-treated water above the filter, serving to ensure the hydrostatic load allowing filtration, is constantly aerated by process air bubbles rising to the surface. In case of the arrival of grease with the influent, this zone serves as a grease removal area and allows floating fats that do not penetrate through the filtering material. These fats can be easily removed by the next wash.

3.3.10.1.2. B2A® process

B2A® is an up-flow biofilter that ensures the biological degradation of pollution in wastewaters by a biomass fixed on a dense support material at multiple granulometries. It also ensures the filtration of waters.

B2A® process can treat raw water that was only subjected to a screening-grit removal pretreatment. Wastewater passes directly into the biofilter, which involves eight layers of materials of various granulometries and densities decreasing from the bottom to the top of the biofilter in the direction of filtration.

The high granulometries of the low part, replacing the primary settling tank, retain the suspended solids in the raw water without leading to a too rapid increase in head loss. The upwards hydraulic flow imposes a controlled operation supported by parts that equally distribute the flow through an originally designed distributing floor and thus minimizes clogging risks. Primary-type SS retention is conducted at the bottom of the biofilter. A large part of the particulate pollution is therefore trapped and hydrolyzed. This surplus of solubilized particulate organic material leads to an improvement in denitrification performances.

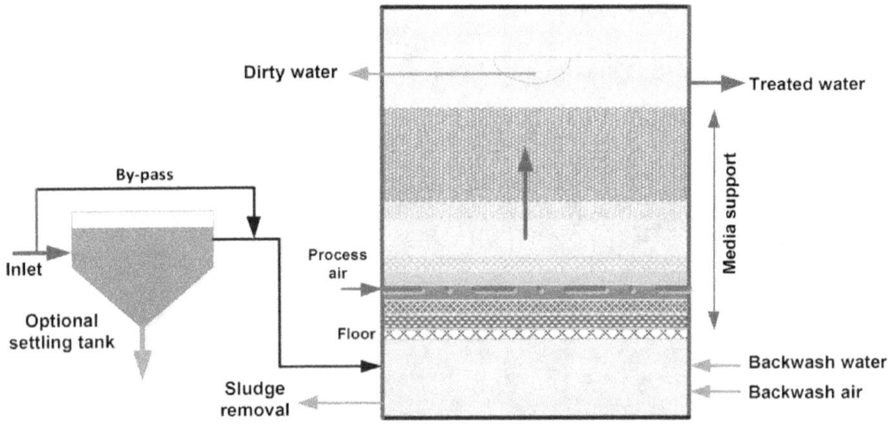

Figure 3.33. *Schematic diagram of B2A®*

Various media supports consist of, from bottom to top, rolled pebbles (40–100 mm), crushed gravel (6–40 mm) and biodagene (2–7 mm). Including all layers, the height of material ranges between 3.5 m and 4.5 m.

The distribution floor is the combination of a material-supporting part, pierced with large holes that do not facilitate clogging, and the layer of pebbles for the hydraulic distribution. The concrete floor has special nozzles with large holes (of 11 mm) with a density of 25 units/m^2.

Air is blown through perforated ducts at an intermediary level in the material or in the low part under the distribution floor.

In case of a ramp in the material, this involves 100 orifices/m^2 of filter. In case of a ramp under the floor, this involves as many orifices as there are special nozzles on the floor, which means 25 holes/m^2 of filter.

The process generates two types of sludge: a primary sludge at the basis of the filter and a biological sludge attached to the upper layers. Washing has therefore two phases:

– The first phase is counter-current flow to the process by deep purge. It requires a velocity in water of 20–40 m.h^{-1}. It allows the removal of primary sludge of lower layers with high granulometry.

– The second phase uses backwash air and water to clean the upper layers. Air velocity is 50 m.h^{-1}. A water velocity of 30–40 m.h^{-1} is sufficient to backwash the material, but a velocity at the end of washing of 60–80 m.h^{-1} is required to preserve the classification of layers of material.

The surface of a cell is maximum 84 m^2.

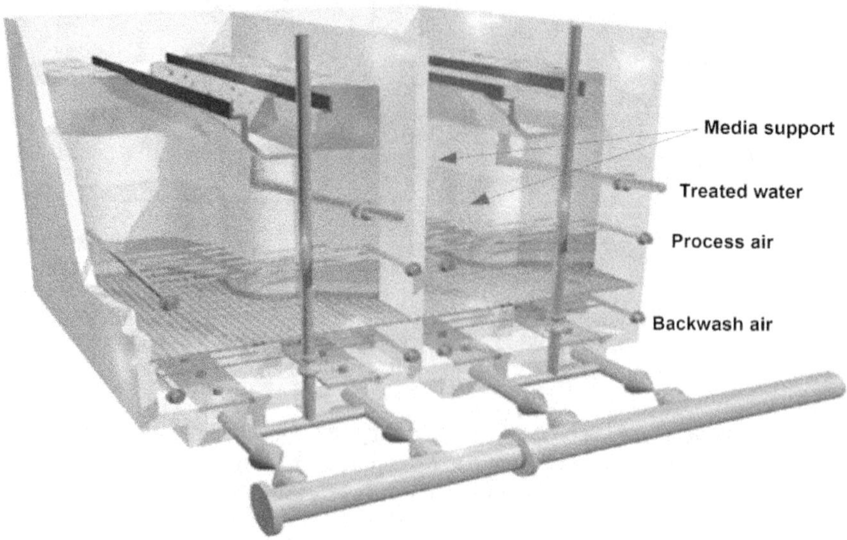

Figure 3.34. *Basic section of B2A®*

3.3.10.1.3. Biofor® process

Biofor® is a biofiltration process on an immersed dense media support, developed and commercialized by Degrémont (Suez). Water to be treated is supplied up-flow in co-current flow with the process air injected at the base of the filtering bed by an aeration ramp. The reactor is supported by a material ensuring the fixation of a purifying biomass and the retention of TSS by filtration (Degrémont 2017).

The material is expanded clay known as biolite with variable granulometry, depending on applications, from 3 to 6 mm. With a density above 1.2, biolite offers a high roughness and a wide specific surface. This characteristic allows the preservation of a bacterial film sufficient to restart the biological process very rapidly, even after a vigorous air and water wash. However, the reactor might be supported by other types of materials. Indeed, three types of materials are commonly used for aerated Biofor®:

– L 2.7 (crushed) biolite or P 2.7 (round) biolite, of effective size ranging between 2.5 and 2.9 mm;

– P 3.5 biolite, of effective size ranging between 3.2 and 3.8 mm.

Other granulometries can be chosen for specific applications. The height of the material can be up to 5 m. In order to prevent the clogging of the parts distributing the water to be treated, the biofilter is generally preceded by a fine screening, which may reach 3 mm spacing.

Biofor® operates in water and air up-flow, which has many advantages: water passage velocity may be significant, high SS retention capacity (proper SS spreading through the filter depth), high oxygenation efficiency (Degrémont 2017).

Oxygenation is conducted by process air diffusion at the base of the filtering material. Equal distribution is ensured by a set of mesh collectors equipped with diffusers with Oxazur membranes.

Air and water periodic backwashes allow the removal of sludge accumulated in the material due to the retention of SS and the bacteria proliferation. These air and water backwashes are automated and initiated by a maximal head loss or by a predefined maximal duration of the cycle. Dirty backwash water is recovered in a tank and then treated and recycled.

Entirely automatic backwashes are conducted by means of air and water. Wash water is generally treated water stored in a tank. Backwashes are conducted at velocities of 20–30 $m.h^{-1}$ in water and 70 $m.h^{-1}$ in air. The volume of wash water is 8 to 10 m^3 of water per m^2 of filtration, or 2.8 to 3.5 m^3 per m^3 of washed material.

Backwashes are conducted at co-current flow with the supply, which requires SS being preferentially retained at the bottom of the filter to cross the entire filter. The backwash of the filter at co-current flow for the supply requires good control of the fluidization of the material in order to carry SS and remove them efficiently. Dirty waters are removed by the top of the filter.

Developed at the beginning of the 1980s for the secondary treatment of urban wastewaters, Biofor® was gradually implemented in several types fulfilling various functions. They can be regrouped in two families:

– Aerated Biofor®: Biofor® C for carbon removal, Biofor® CN for carbon removal and partial nitrification, Biofor® N for advanced nitrification;

– Non-aerated Biofor®: pre-DN or post-DN Biofor® for denitrification.

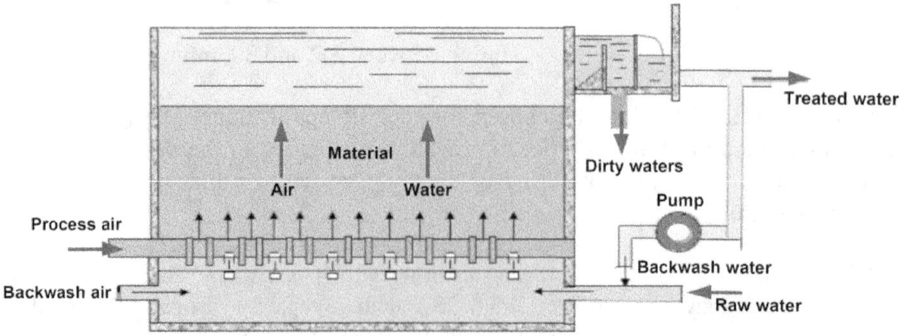

Figure 3.35. *Schematic diagram of Biofor® process*

A Biofor® facility involves a group of identical reactors generally consisting of a rectangular concrete tank and involving each of the following (Figure 3.35): a water-to-be-treated supply well, equipped with a protection screen; a floor supporting granular materials with a system of fluid introduction and distribution; two frontal spillways for taking-off treated water and washing water – these spillways are protected by a spilling picket fence removing turbulences and releasing air bubbles possibly attached to the biolite grains; a front collection duct for each filter; and a part of the collection channel common to the group of filters.

The floor supports two intermediary support layers and 3 to about 4 m of biolite material. The equal distribution of fluids introduced under the filter floor (raw water, washing water, washing air) is ensured by nozzles of the same type as those of sand filters but especially adapted to residual waters, at a rate of about 50 per m^2 (Degrémont 2017).

The introduction and equal distribution of fluids introduced under the filter floor (raw water, washing water, washing air) are ensured by the process air conducted in the support layer by diffusers with deformable elastic membrane, at a rate of 25 or 50 units per m^2 of floor.

Implementation

Biofor® is normally implemented after primary settling, physico-chemical treatment or biological treatment for treating urban or industrial residual waters.

Its preferred domains of use are the following:

– applications where place or soil quality constraints mean a significant advantage for its compactness (three to five times less ground surface than activated sludge);

– plants with heavy hydraulic overload (e.g. rainwater discharge). With biomass being fixed, it cannot be leached, and the performances, including those of nitrification, are restored not later than after the first wash.

Performances

Types of Biofor®	Applications	Parameters	Values
Biofor® C	Carbon (BOD)	Water velocity ($m^3.m^{-2}.h^{-1}$) Applied loads (kg $BOD.m^{-3}.d^{-1}$)	3–16 2–4
Biofor® CN	Carbon and partial nitrification	Water velocity ($m^3.m^{-2}.h^{-1}$) Applied loads (kg $BOD.m^{-3}.d^{-1}$) Nitrified load (kg $N\text{-}NH_4.m^{-3}.\ d^{-1}$)	3–12 1.2–2 0.4–0.6
Biofor® N	Tertiary nitrification	Water velocity ($m^3.m^{-2}.h^{-1}$) Nitrified load (kg $N\text{-}NH_4.m^{-3}.\ d^{-1}$)	3–12 1.2–1.6

Table 3.10. *Velocities and loads admissible in aerated Biofor® system (T °C from 15 to 20°C)*

Applications	C	C & N	N
Upstream treatment	Primary settling		Biological treatment
Biolite	P 3.5	L2.7	L2.7
SS	60–80%	65–85%	40–75%
BOD	65–85%	70–90%	40–75%
COD	55–75%	60–80%	30–60%
Nitrification	–	< 75%	80–95%

Table 3.11. *Typical removal efficiencies in aerated Biofor® system*

Tables 3.10 and 3.11 indicate velocities and loads obtained on various types of aerated Biofor® systems, as well as common average performances (Degrémont 2017). These values are susceptible to varying as a function of various conditions: temperature, biodegradability and typology of water to be treated, particularly with regard to soluble/particulate ratios.

On primary settled urban residual waters with average concentration, volume loads from 2 to 4 kg $BOD.m^{-3}.d^{-1}$ can be used to reliably obtain a BOD below

25 $mg.L^{-1}$. They can even exceed 6 kg $BOD.m^{-3}.d^{-1}$ with diluted waters (BOD below 70 $mg.L^{-1}$). These applied organic loading rates vary depending on the temperature. In contrast, the granulometry of the material used has little effect on BOD removal.

The maximal nitrification capacity (in kg $N\text{-}NH_4.m^{-3}.d^{-1}$) is strongly influenced by the temperature, as any nitrification process, and also by the ratio of readily absorbable carbon/$N\text{-}NH_4^+$ and therefore by the upstream treatment.

When a partial nitrification is required (efficiency in KN below 70%), a single stage of biofiltration is sufficient; then we have Biofor® CN. The recommendation is to remove a maximum amount of carbon upstream through a primary physico-chemical treatment.

When advanced nitrification is the objective (NK efficiency from 80 to 96%), it is advantageous to remove the maximum amount of carbon in the first stage. Le Biofor® N will then be preferably preceded by a Biofor® C stage or by an average load of activated sludge treatment. Nitrification of about 1.6 kg $N\text{-}NH_4.m^{-3}.\ d^{-1}$ at a temperature of 20°C is then possible.

For intermediary nitrification (65–85%), it is recommended to compare the solutions in a single stage (Biofor® CN) or in two stages (Biofor® C followed by Biofor® N) before making a decision.

Biofor® performances depend on the choice of material, on the SS concentration of the influent and on the hydraulic load in filtration.

For treating water presenting an initial SS content below 100 $mg.L^{-1}$, the efficiencies of SS removal vary from 75 to 85% for average water velocities ranging from 3 to 6 $m.h^{-1}$.

Excess sludge production

Excess sludge production is the consequence of SS retention and of the growth of heterotrophic and autotrophic biomasses. It is characteristic for average-load biological systems, taking into account better trapping of SS. The SS retention capacity of a biofilter between two washes normally ranges between 2 and 4 kg $SS.m^{-3}$ of material.

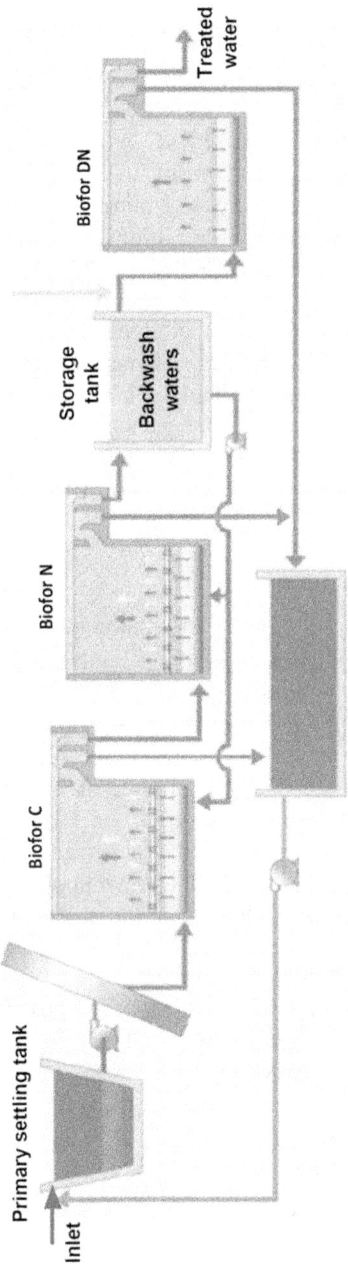

Figure 3.36. *Carbon and nitrogen treatment system with Biofor®*

These SS are recovered in the form of washing waters whose average concentration varies between 200 and 800 $mg.L^{-1}$ depending on the application. They are either returned upstream to the primary treatment or treated separately by settling or flotation.

Velocity of process air

The applied velocity of process air ($Nm^3.m^{-2}.h^{-1}$) must facilitate meeting the following oxygen needs: carbon oxidation, N-NH_4 nitrogen oxidation and endogenous respiration. According to the application conditions, it can vary between 5 and 25 $Nm^3.m^{-2}.h^{-1}$, and even more in tertiary nitrification at high temperature.

The oxygen transfer efficiency, under actual conditions, ranges from 15 to 25%; the oxygenation energy consumption ranges between 0.5 and 0.7 kWh per kg of removed BOD (at full load) or between 2.5 and 3.1 kWh per kg of nitrified N-NH_4.

In addition to this expense, there is the energy related to the periodical washing of material, about 0.1 kWh per kg of BOD removed in Biofor® C and 0.15 kWh per kg of N-NH_4 nitrified in Biofor® N.

Non-aerated Biofor® (denitrification)

Non-aerated Biofor® DN can be used for the biological reduction of nitrates into molecular nitrogen by denitrifying bacteria fixed on the material. The design of Biofor® DN resembles that of aerated Biofor®. The difference resides in the elimination of the process air supply network and the granulometry of the material.

Biofor® DN uses essentially the P 4.5 biolite, of an effective size ranging between 4.2 and 5 mm, without a support layer. P 3.5 biolite can also be used in certain applications. The effective size of the support material to be used in denitrification is always larger than the one used in other applications. This is explained by the adhesive nature of denitrification sludge and the necessity of vigorous washes.

Depending on their position in the treatment system, there are two types of Biofor® DN: Biofor® pre-DN and Biofor® post-DN. The absorbable organic carbon required for denitrification is brought by the rapidly biodegradable BOD of the primary settled water in the first case or by a source of external carbon, generally methanol, in the second case.

Examples of systems using Biofor® DN

The criteria for the application and dimensioning of the two types of Biofor® DN are significantly different (Table 3.12).

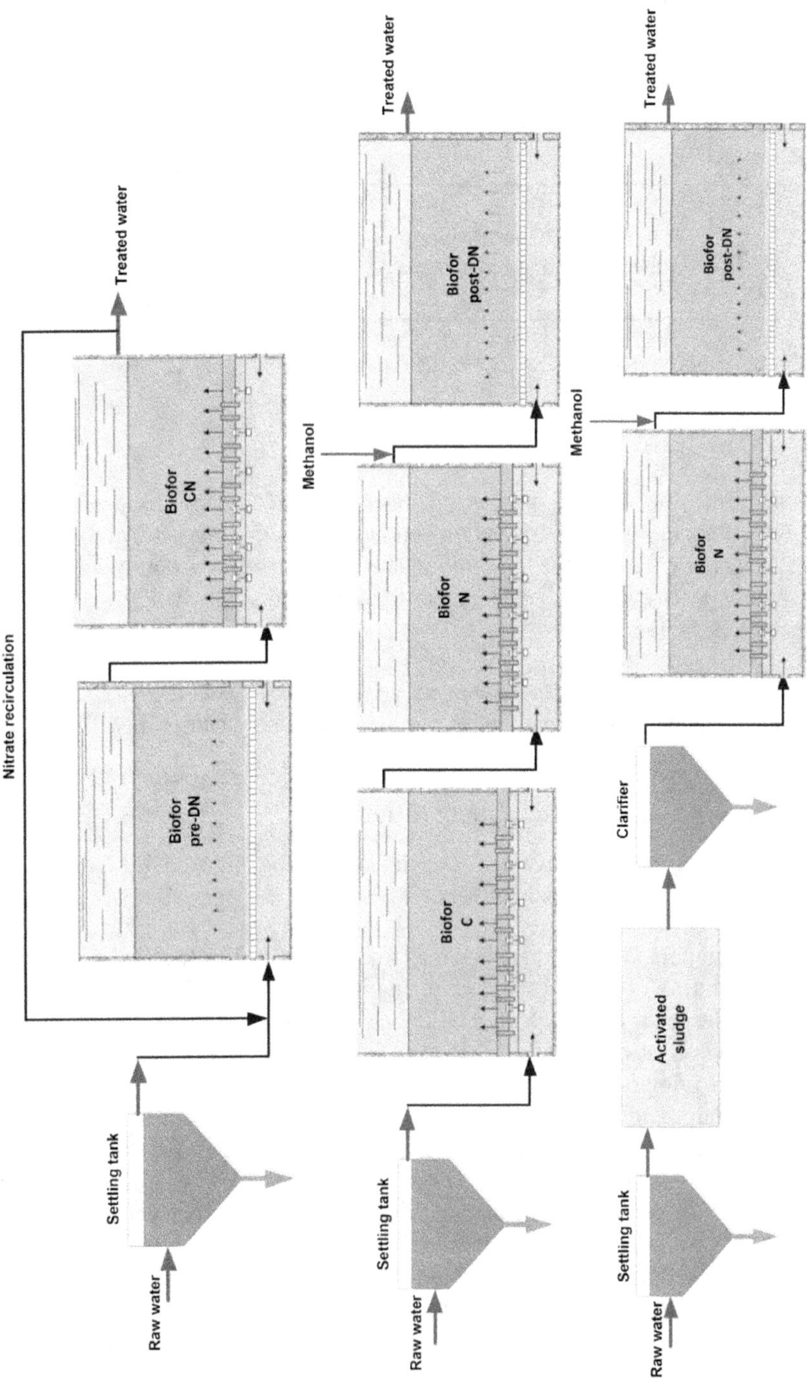

Figure 3.37. *Examples of Biofor® DN applications*

Biofor® type	Application	Parameters	Values
Biofor® pre-DN	Upstream denitrification	Water velocity ($m^3.m^{-2}.h^{-1}$)	10–30
		Denitrified load (kg $N\text{-}NO_3.m^{-3}.d^{-1}$)	1–1.2
Biofor® post-DN	Downstream denitrification (with an addition of methanol)	Water velocity ($m^3.m^{-2}.h^{-1}$)	10–35
		Denitrified load (kg $N\text{-}NO_3.m^{-3}.d^{-1}$)	25

Table 3.12. *Acceptable velocities and loads in Biofor® DN (T°C from 15 to 20°C)*

Advantages

In terms of the biofiltration process, Biofor® has advantages related to this technique. Filtration at co-flow with aeration and in the direction of material expansion not only prevents problems of filter embolism but also allows applying higher velocities. Water at the surface of filters is treated water, which allows limiting odors in secondary treatment.

The process is applicable to the treatment of both urban and industrial residual waters, being part of many applications in the treatment of paper mill effluents (a dozen), agrifood effluents and the chemical industry.

3.3.10.1.4. Biodrof® process Degrémont (Suez)

This is a trickling bed biofilter also developed starting in 1985 by Degrémont (Suez) in cooperation with a German university.

The media support of the reactor is a fine material (biolite) allowing the treatment and filtration of wastewaters. Oxygenation is conducted by air aspiration (at co-flow with the feed) by means of ventilation. Washing is conducted by means of air and (treated) water at counter-flow with the feed. Biofiltration is down-flow. The water issued from primary settling tanks is directed to the biofilter. Biofiltration is aerated by a vacuum pump system under the biofilter that allows maintaining a water-air co-flow in the gravel pack from 1.6 to 3 m height. The biofilter is never immersed (except for washes); this is why they are referred to as dry biofilters. Treated water crosses three layers of gravel, each with 7 cm thickness and different granulometries. These gravel layers are there to prevent filtering media from escaping towards the retention tank and damaging the removal pump. Water passes through the nozzles fixed in the concrete slabs supported at their turn by concrete pillars at the four corners of each slab. Water is then collected in the lower part of the biofiltration modules, directed to the retention tank, and then removed by means of a pump to the receiving medium. The drain pipe is submerged in water to form hydraulic protection and maintain the bottom of the biofilter under low pressure (Degrémont 2017).

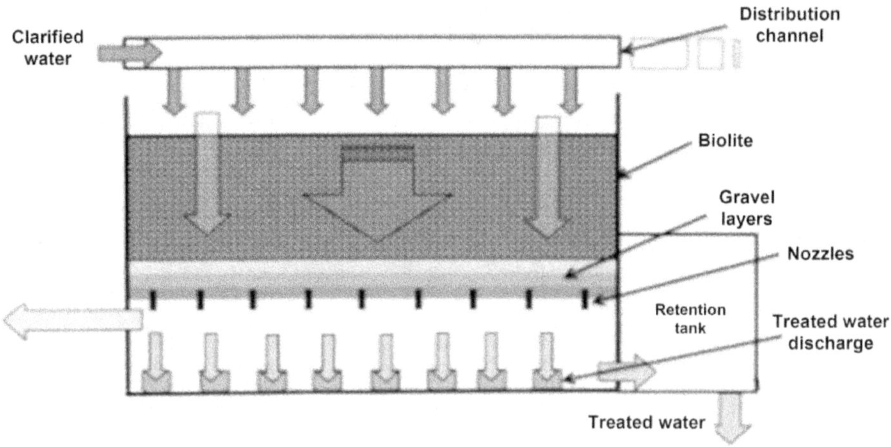

Figure 3.38. *Operating diagram of the Biodorf® process*

The backwash system of the biofilter is automated, but human intervention is possible, particularly for preparing the system for an expected rain event with a high flow rate. Backwash protocol has three different stages: backwash, mini-backwash and declogging. These stages follow an automatically controlled protocol:

– Backwash and mini-backwash: their operation is similar. The inlet and outlet of the biofilter are closed by the use of guillotine valves. The inlet of the backwash water is then open at the level of the retention tank, and the biofiltration module is fully filled. Air is then blown, and at the same time, backwash water continues to be injected and recovered at the top of Biodrof®. Water and air are injected in an up-flow to put the media in suspension and declog the biofilter. The difference between mini-backwashes and backwashes is based on backwash time and flow rates used.

– Declogging: unlike washes and mini-backwashes, declogging is conducted when the biofilter is operating (inlet and outlet of the biofilter remain open). Declogging passes through a stage of filling with backwash water and up-flow air injection. Declogging durations are relatively short.

Biodrof® is appreciated for its simple design, excellent retention of TSS and low energy consumption. This is explained by the fact that this technique has a low aeration capacity, consequently being often used as a TSS polishing tool.

Similar to all processes with up-flow filtering, i.e. raw water is in open air, odors are generated. If clogging is not controlled, the bed is no longer dewatered and aeration becomes insufficient.

3.3.10.1.5. Biolest® (Stereau-Bouygues) process

Biolest® is a process that combines the advantages of biological treatment on a fixed culture and the filtration on granular material: biozzolane. It involves a biological filter immersed in an up-flow on a dense material.

The principle of biofiltration with upward co-flow of water and air on a material denser than water and non-compressible, combined with the intrinsic characteristics of biozzolane, provides a high-velocity operation without compaction of the material and with excellent performances.

The process involves air and water co-current flow filtration, and filter washing is conducted with a counter-current flow of feed using air and water.

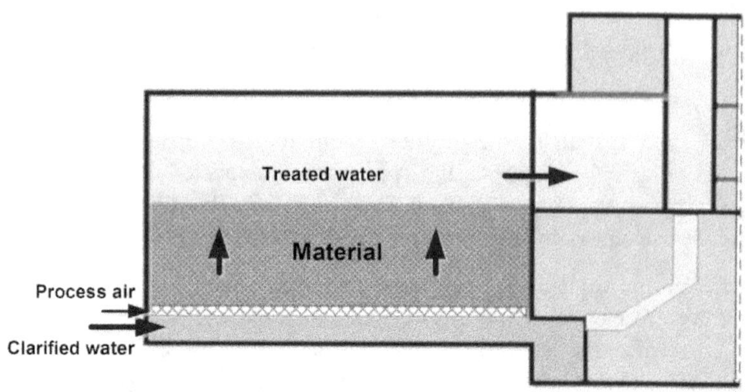

Figure 3.39. *Schematic diagram of Biolest®*

Biozzolane is a natural material resulting from the pouzzolane volcanic rock, which thanks to its porosity and chemical composition has remarkable properties allowing a rapid development and an excellent fixation of bacteria on the material. The Biolest® process can be presented as follows:

– a very compact process;

– natural media: biozzolane;

– perfectly distributed aeration for better transfer in the biomass.

Its field of application extends from carbon removal to post-denitrification for urban and industrial effluents. The Biolest® process is particularly adapted to compact plants with load variation or with diluted effluents that must be integrated in a sensitive environment. This process is applicable to carbon treatment (Biolest® C) and to nitrogen treatment: nitrification (Biolest® N) and denitrification (pre-DN Biolest® or post-DN Biolest®).

3.3.10.1.6. Biopur® process (AFS Aqua AG-Wabag)

Biopur® is a process with fixed cultures involving a reactor immersed with downward flow on an ordered plastic structure having a wide specific surface (about 250 $m^2.m^{-3}$) or on a granular material (expanded clay). The process was developed by the German company Wabag.

Air needs are met by forced aeration (process air); periodic washes are conducted by air and water counter-flows with the feed.

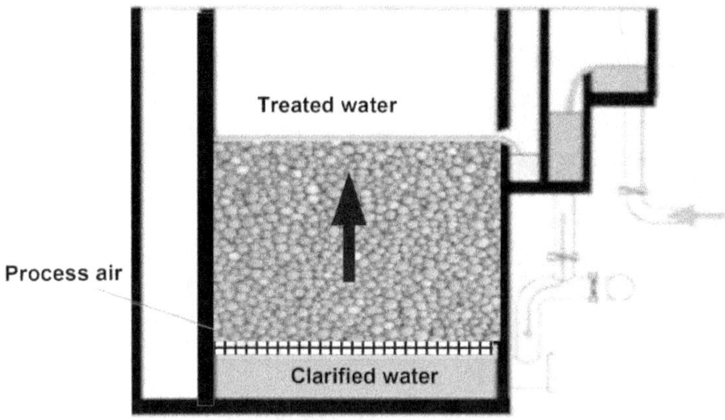

Figure 3.40. *Biopur® CN*

As the structure implemented in Biopur® does not allow very efficient filtration, Biopur® is generally followed by an SS refining stage in a filter known as bioactive, which is a classical two-layer filter (down-flow at counter-flow washing). The main advantage of the process is that the use of this ordered structure as a support material involves very low head losses and therefore avoids intermediary pumping; on the other hand, this low head loss is obtained to the detriment of the filtration effect, therefore requiring a refining treatment.

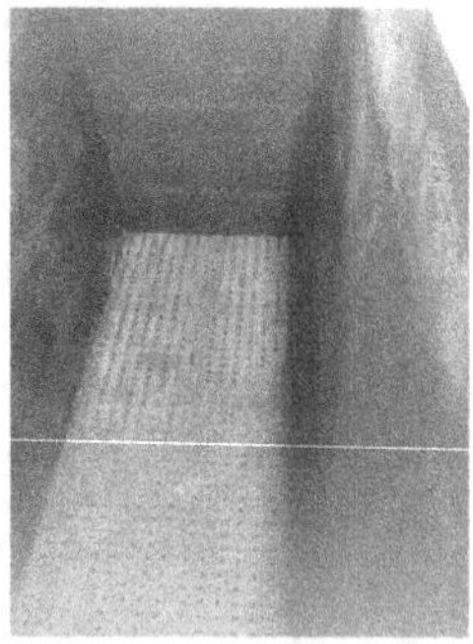

Figure 3.41. *View inside an empty Biopur© CN tank*

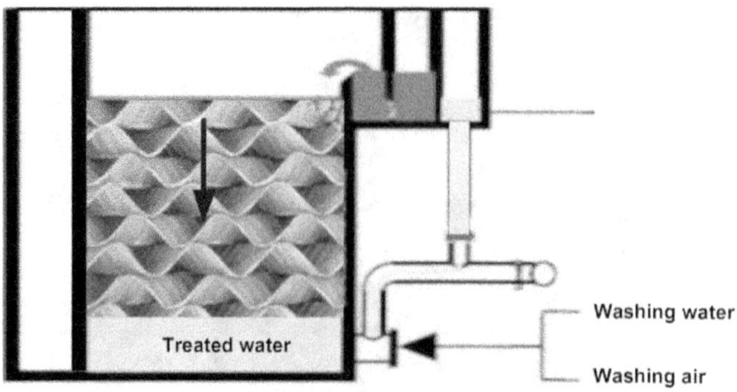

Figure 3.42. *Biopur® DN*

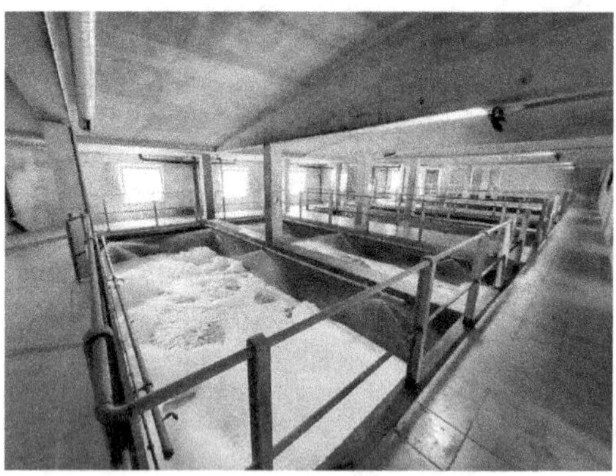

Figure 3.43. *View on Biopur© C*

Due to a significant vacuum percentage (90% against 30% for the Biostyr®), applied velocities can be very important.

Applications of this process cover the treatment of both urban and industrial wastewaters, removing carbon, nitrogen and phosphorus pollution.

Material	**Application**	**Dimensioning**
Ordered material 	Carbon removal (Biopur® C) Pre-denitrification (Biopur® DN) Nitrification (Biopur® N)	Height of material: 3–6 m Specific surface: 250–400 $m^2.m^{-3}$ Filtration velocity: up to 25 $m.h^{-1}$ Cycle duration: 24–72 h
Granular material 	Carbon removal (Biopur® CK) Pre- and post-denitrification (Biopur® DNK) Nitrification (Biopur® NK)	Height of material: 3–4 m Specific surface: 600–1,200 $m^2.m^{-3}$ Filtration velocity: up to 20 $m.h^{-1}$ Cycle duration: 24–48 h
Bioactive clarifying material 	Flocculation-filtration Post-denitrification Removal of residual ammonia and nitrites Monolayer or multilayer	Height of material: 1.2–2.5 m Filtration velocity: up to 20 $m.h^{-1}$ Cycle duration: 24–48 h

Table 3.13. *Characteristics of supports and materials used in Biopur® process*

Each parameter of the pollution is removed on a unit stage; denitrification could be conducted downstream or upstream of nitrification (with recycling in this case).

The external carbon source mentioned in the case of post-denitrification is the acetate.

The applied loads and velocities are summarized in Table 3.14.

Type	Function	Loads, velocities and efficiency
Biopur® C	Carbon removal	Cv = 10 kg DCO.m^{-3}.d^{-1} V = 20 m.h^{-1} BOD5 efficiency ≈ 85%
Biopur® N	Nitrification	Cv = 0.2–0.5 kgN.NH$_4$.m^{-3}.d^{-1} Outlet N.NH$_4$ < 1 mg.L^{-1} Cv = 3 kg DCO.m^{-3}.d^{-1}
Biopur® DN	Upstream denitrification	0.4 kg N.NO$_3$.m^{-3}.d^{-1} Outlet N.NO$_3$ < 1 mg.L^{-1}
Bioactif	SS refining, post-denitrification, precipitation of phosphorus	Cv = 1 kg N.NO$_3$.m^{-3}.d^{-1} V = 5–8 m.h^{-1} Outlet SS < 5 mg.L^{-1} Outlet Ptotal < 0.2 mg.L^{-1}

Table 3.14. *Loads and velocities applied for various Biopur® configurations*

3.3.10.2. *Moving bed processes*

3.3.10.2.1. Biostyr®

Biostyr® is a biofilter with upward flow that ensures the biological degradation of pollution by a biomass fixed on the floating support material and simultaneously the filtration of these waters. This process implements a floating synthetic granular support in the form of expanded polystyrene beads. Bacteria attach to the granular material; in order to develop, they use components dissolved in the water to be treated and also the oxygen of the blown air. Pollution is degraded and transformed into cellular material that is trapped in the filtering bed by mechanical retention (Gilles 2004).

Particularly compact, Biostyr® associates in a single structure the biological reactor where pollution is transformed and the separator of phases where the treated water is freed from the materials produced by the treatment. Air is blown either at the bottom of the bed or inside the material. In this latter configuration, the filter can simultaneously perform nitrification and denitrification functions. It removes all biodegradable pollutants: carbon pollution (COD and BOD), suspended solids (SS), ammonia (N-NH$_4$) and nitrates (N-NO$_3$).

Unlike the up-flow filters on materials denser than water, the filter head loss ensures the equal distribution of the effluent to be treated without using nozzles or any other device susceptible to clogging and without the need to screen the water to be treated. Filtration is conducted in the direction of compaction and not of the expansion of grains, which leads to excellent capture of suspended materials. The periodic counter-flow washing removes the sludge produced and the filtered materials, without them crossing throughout the bed. The downward scouring removes the residues by the shortest path, in the direction of the particle fall.

Biostyrene material is retained by a slab featuring nozzles on the opposite side of the feed, which is in contact only with the treated water and easily accessible. Biostyrene is a fine and regular material that provides a very large specific surface. Expanded polystyrene used as support material of the biofilter proved to be the most efficient support for attaching the autotrophic biomass and therefore for the nitrification process. At an identical average granulometry, it can be used to nitrify twice the loads obtained with biocarbon.

The characteristics of the biostyrene material are summarized as follows:

– Due to its synthetic character, the characteristics of the material (granulometry, density, etc.) can be better controlled during manufacturing.

– Given its density below 1, it is a floating material, allowing the development of an anoxic zone in the system and also authorizing less restrictive washing procedures.

– The specific surface of the material being high, Biostyr® has a high treatment potential.

– The controlled granulometry of the material and filtration in the direction of material compaction allow a low SS residual.

The air draught depends on the objectives of the treatment: at the bottom of the biofilter or within the material. Air is often used specifically to remove total nitrogen (TN) in a single reactor through the implementation of an anoxic zone and an aerated zone within the same cell (Figure 3.44). Water to be treated circulates upwards; it crosses successively through the anoxic zone and then the aerated zone separated from the first one by the process air ramp.

A ceiling of slatted slabs ensures the maintenance of the material. Washing water is stored below the filters, and sludge is removed by simple hydrostatic scouring.

The recycling flow rate of the filtered water to bring the nitrates to the level of the anoxic zone varies between 100 and 400% of the treated water flow rate. This process can also be used in tertiary nitrification; the backwash air ramp also serves as a process air ramp.

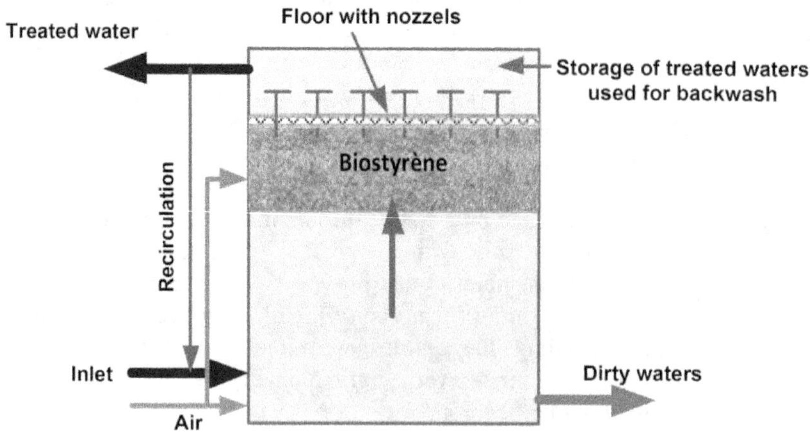

Figure 3.44. *Schematic diagram of Biostyr®*

Performing the nitrification and denitrification functions ensures the removal of all biodegradable pollutants.

The oxygen needs of the aerobic biomass are met by air draughts. This is conducted by perforated ramps. In a co-flow operation, they can be located in the lower part of the filter or at an intermediate level. This second possibility then allows protecting an anoxic zone favorable to denitrification inside the same biofilter, this zone being in an environment of non-treated water. Air and water flow in the same direction avoids the gaseous embolism phenomenon specific to counter-flow air and water systems.

The range of Biostyr® surfaces extends from 2 to 230 m^2. Small-size standards are available in the form of stainless steel units.

Suspended solids and excess biomass are separated by upwards filtration of the flow through the bed of Biostyrene beads, a material of low density and small granulometry. Beads are immersed and retained by a ceiling with nozzles. Generally speaking, the higher the discharge level, the lower the granulometry of the material. The removal of the biodegradable carbon pollution and of suspended solids (TSS) is conducted in the Biostyr® cell in an aerated medium. For nitrification, the design of this configuration is similar to that of carbon removal. It removes the biodegradable carbon pollution, ammonia and TSS in a single stage in the same cell in an aerated medium – as done by the Biocarbone® process. In the secondary treatment, after simple settling, physico-chemical settling or activated sludge working with a very high organic loading rate, ammonia nitrogen and carbon pollution are simultaneously removed. In tertiary treatment, after an average load of activated sludge, a biofilter or a trickling filter, ammonia nitrogen is mostly removed, with complementary refining of suspended solids and residual carbon.

For nitrification/denitrification, after simple settling or physico-chemical treatment, the removal of carbon pollution, nitrification and denitrification are conducted in the same cell with the recirculation of formed nitrates. Indeed, this is possible thanks to two aerated/non-aerated distinct zones available in the Biostyr® configurations: the nitrates formed during nitrification in the aerated zone are recycled and transformed in the anoxic zone into gaseous nitrogen, using the clarified water as a source of carbon for the reaction.

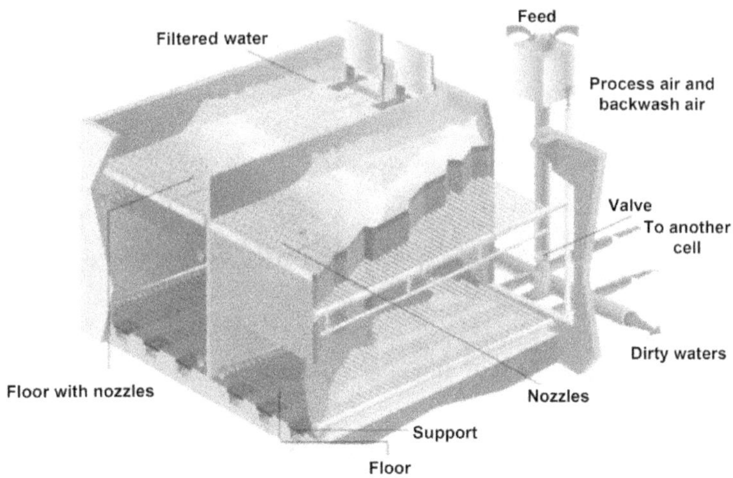

Figure 3.45. *Main Biostyr® equipment*

Thanks to strict control of the oxygen dissolved in the system and smooth control of the injected process air, aerated/non-aerated micro-zones are created around the beads. Nitrification and denitrification processes unfold simultaneously at the level of each bead. Due to this configuration, the recirculation flow rate is reduced by about 25%.

Another configuration of this process allows treating nitrates in a post-denitrification stage. In the final treatment, after the activated sludge, a nitrifying or nitrifying/denitrifying biofilter can be installed. Post-denitrification is conducted in a non-aerated medium when nitrogen removal on the upstream stage is insufficient for meeting the required level of TN. With the contribution of an external carbon source (methanol), the denitrification process allows the transformation of excess residual nitrate into gaseous nitrogen, with very high denitrification efficiency.

Biostyrene material facilitates backwashing due to its synthetic and light structure. Its size and density can be easily adapted to the set objectives. The fact that the process combines reactor and filter, without additional clarification, allows advanced treatment in extremely compact structures with a residence time of 1 to 2 hours of secondary treatment, sufficient for obtaining a discharge compliant with the strictest standards.

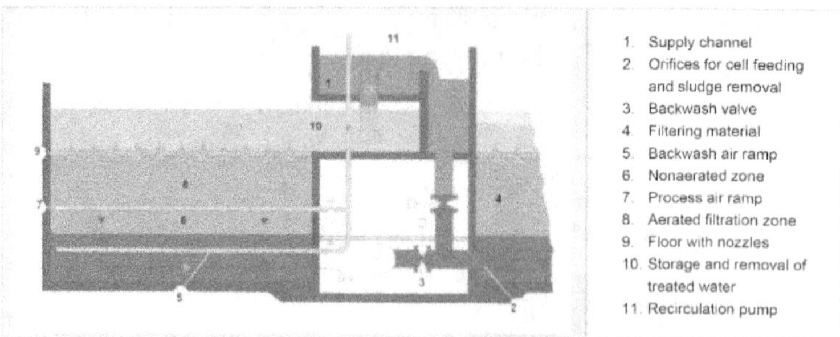

Figure 3.46. *Biostyr® with the two aerated and non-aerated zones*

The combination of an upward liquid flow and of a floating material associates the advantages of downwards flow filters and upwards flow filters, which offers the following possibilities:

– introducing the backwash reserve in the top part of the filter without a separate storage tank and very easily access the nozzles without drawing out the material;

– installing an aeration ramp in the biofilter medium in order to separate a non-aerated zone at the bottom and an aerated zone at the top of the biofilter if nitrification–denitrification must be ensured;

– minimizing odor and aerosol emissions: the ambient air is in contact only with oxygen-saturated treated water; water stripping from pollution-loaded supply is excluded;

– dirty waters remain in a closed chamber without exposure to the atmosphere.

General characteristics of Biostyr® process

– Compactness allowing easy integration.

– Upwards filtration allowing wastewater confinement at the bottom of the structure and presenting treated water at the surface.

– Counter-current flow washing conducted without gravity pumping.

– Filtering material consisting of fine polystyrene bubbles is not biodegradable and does not require renewal in time.

– Possibility to remove carbon and nitrogen pollution from the same structure.

– High concentration of biomass fixed on the support material.

– Efficiency proven during high load and flow rate variations.

– High-performance filtration.

– No final clarification.

– Filtration in the direction of compaction and not of grain expansion, which leads to excellent trapping of suspended materials.

– Filtration velocity between 6 and 30 $m.h^{-1}$ depending on applications.

– Automated biofilter management system.

– Upward flow on a floating material.

– Access to nozzles without removing the material.

– Protection of parts sensitive to clogging (nozzles in contact with treated water).

– Backwash water reserve at the top of filters without a separate storage tank.

– Modularity of the aeration ramp depending on applications.

– Counter-current flow washing by “gravity” scouring without needing a pump for the treated water.

– Sequenced air injections.

– Automated washing.

– Fewer nuisances.

– Ambient air is only in contact with treated water.

– Backwash water, directly collected at the bottom of biofilters, is never in contact with the atmosphere.

Implementation

Due to the evolution in the manufacturing of material, it is possible to increase its granulometry and density. Available average granulometries are 3.2–3.6, 4, 4.5 and 5 mm. The highest granulometries allow the application of carbon pollution removal and post-denitrification.

Due to a fine control of process air, it was possible to design a new operation in nitrification–denitrification known as simultaneous nitrification–denitrification.

The height of the material ranges between 1.5 and 4 m.

The application of aeration ranging between 15 and 30 Nm^3 $air.kg^{-1}$ of applied BOD_5 is sufficient to ensure maximal removal of the carbon pollution. The increase in aeration does not improve the reduction in carbon pollution but induces a stronger nitrification process (Rocher 2006, 2008). However, the aeration intensities to be applied to maintain maximal treatment performances depend on the conditions of air bubble diffusion through the support material (diffuser type, height and nature of the material) and are therefore susceptible to slight variation from one facility to another.

Efficient backwash, similar to Biocarbone®, remains essential for the proper operation of the system. Backwash is conducted by a sequence of air and water phases. Water used for backwash is treated water stored above the ceiling. The required volume of water is 2.5 to 3.5 times the volume of material. The storage of treated water for backwash involves a common reserve above all filters in place. Therefore, there is no need for a dedicated tank of treated water; it is integrated in to the biofilters.

The backwash water phase results from gravity scouring (opening of the foul water removal valve). Therefore, one of the advantages of Biostyr® is the absence of pumping for backwash water phases. The correct washing of the entire filtering material requires backwash air to be provided by an air ramp at the bottom of the filter.

Biostyr® filter washing is also the object of full automation, which is essential to the proper operation of the system.

Biostyr® in the removal of only carbon pollution

The design with aeration at the bottom of the filter allows a material of 4.5 or 5 mm, of average size. The irreversible clogging problem is shifted towards zones with higher applied load, and the purifying potential of the material for the carbon pollution is better used. The use of a material with higher granulometry leads to a lower quality SS discharge.

Biostyr® in simultaneous nitrification–denitrification

The design is based on two distinct (anoxic/aerated) zones with recycling of the aerated stage in the anoxic zone. Since anoxic and aerobic zones are no longer distinct, the process air ramp is at the bottom of the filter (air transfer efficiency is improved), and it also serves for supplying washing air. The upper ramp is therefore eliminated. Water recirculation is preserved to return nitrates to the bottom of the filter. The required recycling rate is, however, reduced compared to the classical nitrification–denitrification. A fine control of process air such as cascade regulation, based on continuous measurement of NH_4-O_2 dissolved in water at the filter outlet, allows a supply of oxygen strictly required for efficient nitrification but sufficiently reduced to preserve denitrifying potential. System design is simplified compared to the operation in two distinct filters. Anoxic and aerobic zones are positioned in two stacked layers, each of them having the total biofiltration surface. The applied hydraulic loads (velocities) are reduced, which makes this parameter less limiting than in two distinct filters. Simultaneous nitrification–denitrification that reduces the necessary recycling rate lowers even further the hydraulics in the system.

Biostyr® in post-denitrification

Denitrification is a heterotrophic process of carbon pollution degradation, brought into post-denitrification by an external source. The choice of higher granulometries (4.5–5 mm) allows better use of the purifying potential. The biofilter is non-aerated with a contribution from an external carbon source. The external carbon source must be a rapidly degradable nitrogen-free substrate producing little biological sludge. The only substrate actually usable is methanol.

In the case of a final discharge level requiring only partial denitrification efficiencies (residual N.NO_3 > 3 mg.L^{-1}), to limit the hydraulic loads on the structure and especially limit the presence of nitrites (very toxic N.NO_2) in the treated effluent, it is recommended to treat, with a low N.NO_3 residual, only one part of the discharge in post-denitrification and mix with the non-treated fraction rather than accepting into the structure the entire discharge with degraded performances.

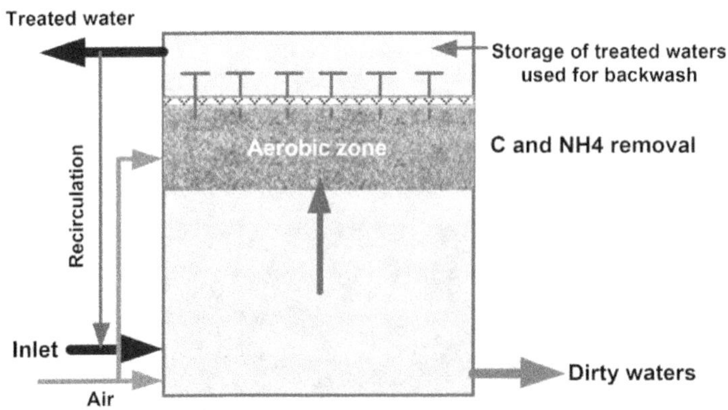

Figure 3.47. *Biostyr®: operating diagram for carbon and ammonia*

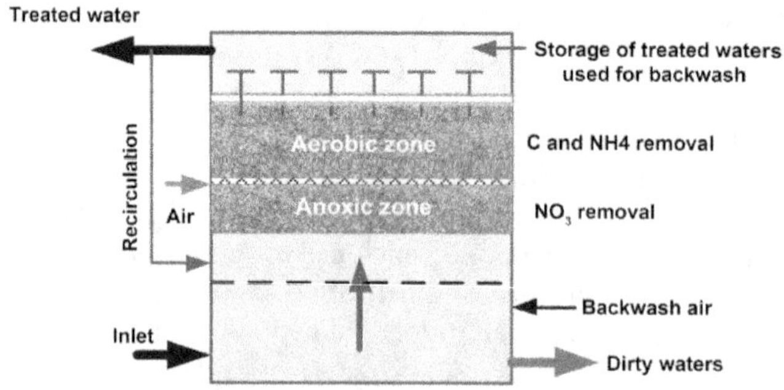

Figure 3.48. *Biostyr®: operation diagram for nitrification–denitrification*

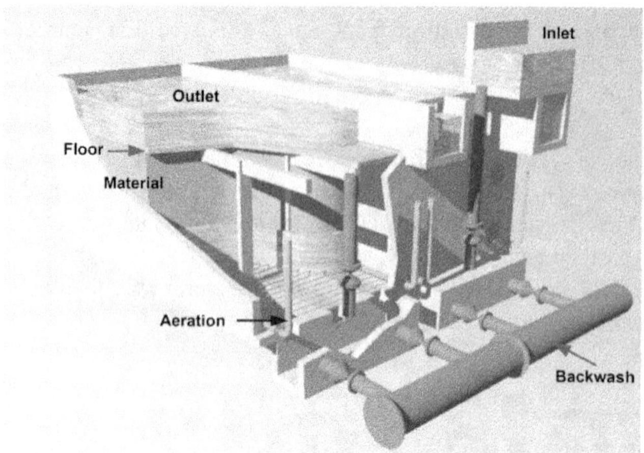

Figure 3.49. *Section view of a Biostyr® cell*

Figure 3.50. *Aeration ramp and slabs*

Figure 3.51. *Biostyr® for washing*

Fields of application

Similar to Biocarbone®, domestic wastewater treatment with Biostyr® requires a primary settling stage. Phosphorus must be treated in another structure of the water row. The fields of application of Biostyr® are therefore:

– removal of suspended materials and oxidizable materials (in secondary treatment);

– removal of carbon and nitrogen pollution (suspended solids, oxidizable organic substances and nitrification or nitrification–denitrification in secondary treatment);

– removal of only nitrogen pollution (tertiary nitrification);

– post-denitrification with addition of external carbon, in a structure downstream of a nitrifying or nitrifying-denitrifying stage of insufficient performance.

Specific Biostyr® advantages

– Filtration in the direction of material compaction.

– Backwash with counter-flow of supply.

– Nozzles for flow rate distribution in a treated water environment, avoiding clogging risks.

– A high nitrification treatment potential.

– Absence of water pumping for backwash.

– Large filter surfaces with up to 230 m^2 of unit surface.

– Backwash simplification.

– The water reserve required to conduct gravity purges, installed above the filter, simplifies the design of the system.

– The absence of risk of embolism (air–water co-current flow).

– Nitrification and denitrification are conducted in the same structure by arranging a specific anoxic zone at the bottom of the filter (for classical Biostyr® N.DN).

– Limitation of generated odors (the air above the filters is in contact with purified water, and sludge is removed without exposure to the atmosphere).

– Easy access to nozzles of the filtering ceiling, making their replacement very easy in case of a problem.

Process performances

	Material (mm)	Carbon pollution	Secondary nitrification	Nitrification–denitrification	Tertiary nitrification
COD loading rate ($kg.m^{-3}$ total.d^{-1})	3.6 4 4.5 5	– – 8 7.2	6 5.4 4.8 –	6 5.4 4.8 –	– – – –
Nitrifiable loading rate ($kg.m^{-3}$ aerated. d^{-1})	3.6 4 4.5	– – –	0.6 0.54 0.48	1 0.9 0.8	1.1 1 0.9
Denitrifiable loading rate (upstream anoxic) (kg. m^{-3} anoxic. d^{-1})	3.6 4 4.5	– – –	– – –	1 0.9 0.8	– – –
Denitrifiable loading rate (downstream anoxic) (kg. m^{-3} anoxic. d^{-1})	4.5 5	– –	– –	1.6 1.4	– –

Table 3.15. *Loading rates applicable at 12°C*

The process's use facilitates:

– reaching a significant treatment capacity: a nitrogen loading rate of 0.8–1 kg $N\text{-}NH_4.m^{-3}.d^{-1}$ can be simultaneously removed with a maximal reduction of COD;

– passage velocities up to 10 $m.h^{-1}$ in tertiary treatment thanks to the co-current air and water flow.

Because of perfectly controlled granulometry and water quality, nitrogen removal efficiencies are improved. The volume loads applied to reach the TN levels (Table 3.15) range between 0.8 and 1 kg of $N\text{-}NH_4.m^{-3}.d^{-1}$.

Table 3.15 gives the order of magnitude of loading rates applicable to reach at 12°C a COD = 80 $mg.L^{-1}$, BOD = 20 $mg.L^{-1}$ for carbon pollution, KTN = 5 $mg.L^{-1}$ for nitrification and $N.N0_3$ = 3 $mg.L^{-1}$ for denitrification.

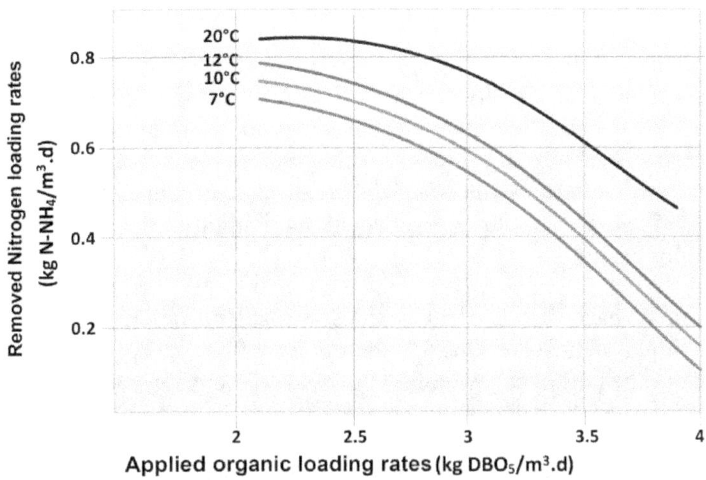

Figure 3.52. *N-NH4 removal efficiencies depending on the applied organic loading rates. For a color version of this figure, see www.iste.co.uk/gaid/wastewaterengineering2.zip*

The physical filtration process generates an accumulation of sludge and causes biofilter clogging. The clogging degree, as a function of treated loads, defines the head losses on the system, the duration of the cycle and the applicable water velocities. Clogging degree, head losses and cycle duration depend on the temperature, granulometry and height of material. High granulometries reduce the clogging level and allow higher hydraulic loads.

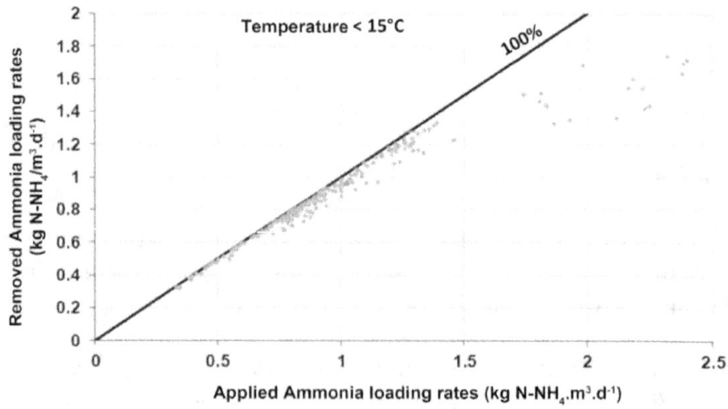

Figure 3.53. *N-NH4 removal with Biostyr®*

The ceiling is perforated by specific nozzles with a density of 49 units/m^2. The air ramp has an aeration density of 100 orifices/m^2 filter with orifices of 1.3 mm. In case of air blown under the material, the air ramp is simplified. It only requires eight orifices/m^2. The size of orifices then ranges between 2.5 and 5 mm (depending on the air flow rates to be blown). The small air cushion that then forms at the interface with the material allows good equal distribution in the filter. This air ramp is also used for backwash.

The "normal" backwash is a regular and systematic operation expected to regularly remove the surplus biomass and the SS retained in the filter. It is preferentially initiated by the clock in the period of low pollution. It is sufficient for stabilizing the head loss after backwash when the polluting load remains moderate.

The mini-backwash procedure is exceptionally initiated in the case of excessive pollutant load or punctual dysfunction of the facility. It allows the partial cleaning of the filter and thus extends the ongoing filtration cycle, waiting for a "normal" backwash. It is automatically initiated when the maximal head loss is reached.

Exceptional backwash is used when, following an exceptional polluting load or an extended dysfunction of the facility, the head loss after backwash drifts beyond the recommended values. Backwash consists of a series of backwash air or backwash water sequences. In contrast to Biocarbone®, there is no air + water phase. Water for backwash is purified water. Velocities in backwash water depend on the type of material (the larger the granulometry, the higher the velocities required for the expansion of the material).

In the main water treatment line, various configurations of this process are possible depending on the required qualities of discharge:

– in secondary treatment for carbon and nitrogen treatment (Figure 3.54);

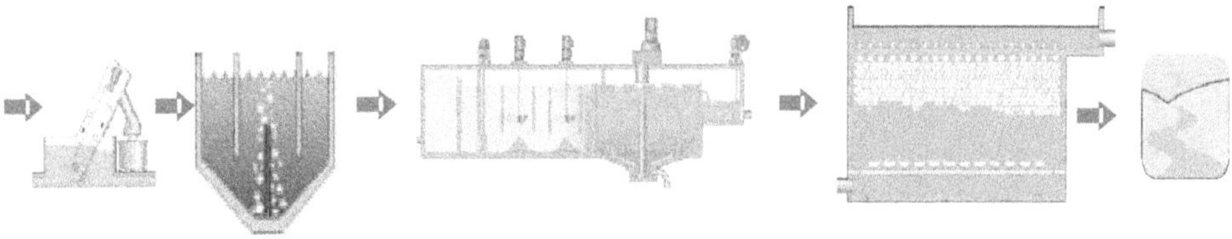

Figure 3.54. *Biostyr® for carbon and nitrogen treatment*

– in tertiary treatment for nitrogen pushed downstream of an existing activated sludge (Figure 3.55);

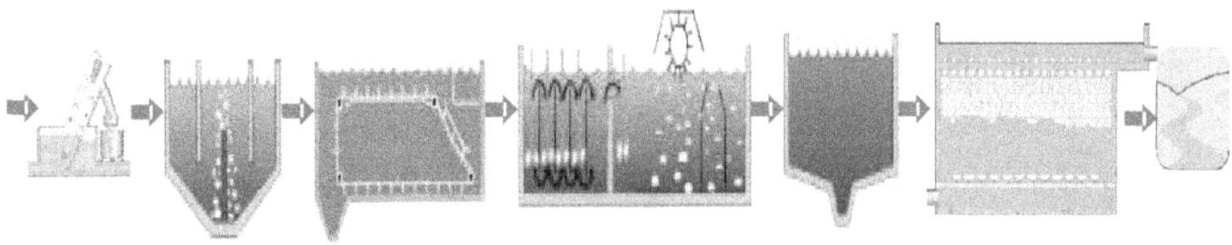

Figure 3.55. *Tertiary Biostyr®*

– in secondary treatment for the advanced treatment of nitrogen (Figure 3.56).

3.3.10.2.2. Biostyr® Duo (OTV-Veolia) process

Biostyr® Duo is a biofiltration solution developed by Veolia Water Technologies to improve the biological performances of the urban wastewater treatment systems. This revolutionary process combines the removal of biodegradable pollutions and physical filtration in a single structure.

Thanks to a patented design, Biostyr® Duo optimizes the water treatment process, offering an efficient solution to meet very strict quality standards.

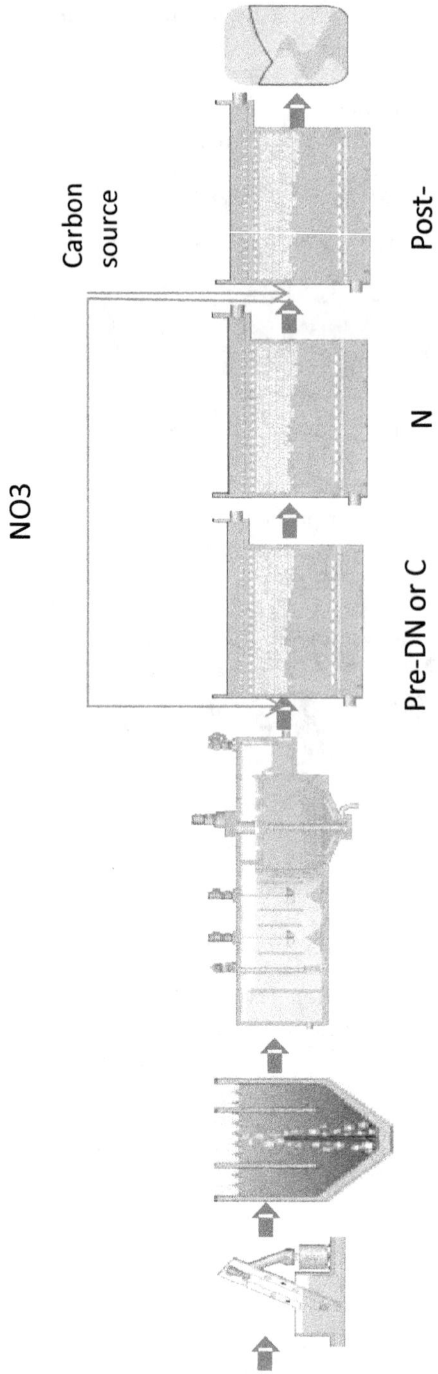

Figure 3.56. *Biostyr® for nitrogen refining treatment*

Biostyr® Duo technology is an upward flow biological reactor whose media support combines two purifying biomass support materials: polystyrene beads and K5 material (AnoxKaldnes™).

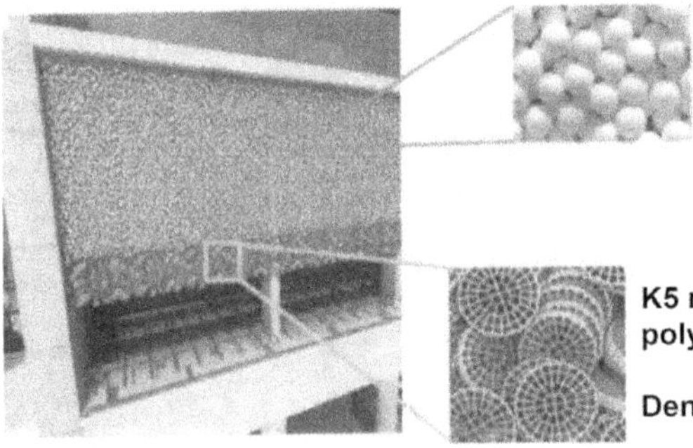

Figure 3.57. *Schematic diagram of Biostyr® Duo*

Biostyr® Duo consists of using the void space under the biostyrene media by adding an additional media layer, in the form of polyethylene chips known as K5, in the same reactor. The result of adding this layer of additional media is:

– reduction of fouling and of backwash frequency;

– reduction of the downstream load on the polystyrene bead media;

– better compactness thanks to more significant removed pollutant loads;

– longer cycle durations thanks to a decrease in head losses on the biostyrene media;

– acceptance of higher SS concentrations;

– protection against biostyrene losses, blocked during backwash by K5 media. Therefore, a bead trap is not necessary, as there is no bead leakage.

Biostyr® Duo operates on the basis of a double-layer biofilter. The first layer consists of biostyrene, a low-density granular material that hosts most of the active biomass. It is used very efficiently in the Biostyr® process.

Under this layer, a second material is integrated to support the additional biomass. Wastewaters pass in upward flow, crossing these two materials successively, which facilitates an advanced biological treatment by reducing the load on the biostyrene. It associates an aerated biological filter and a reactor with biological film with an AnoxKaldnes™ moving bed to treat the soluble and particulate pollution.

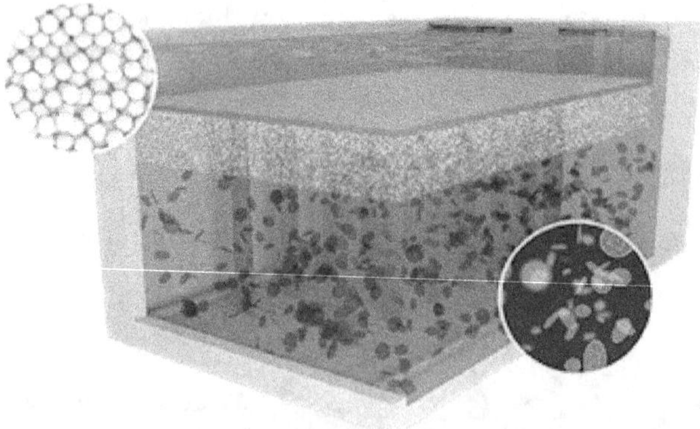

Figure 3.58. *Biological filtration with Biostyr® Duo*

This configuration allows not only etter compactness and more significant removed pollutant loads but also protects biostyrene from losses and premature fouling. The result is extended cycle duration, a reduced washing frequency and an optimized global efficiency.

Biostyr® Duo finds its application mainly in the urban wastewater treatment plants. It is particularly adapted for the intensification of wastewater treatment, without overconsumption of energy and water. Finally, this solution is easily integrated into the existing infrastructures, thus being a sustainable option for the upgrades of current systems. This process is the most compact solution for the removal of carbon and nitrogen. In the case of new facilities, it offers a significant reduction of the covered surface, and for existing facilities, the possibility to increase the unit capacity without major modifications.

This process is also suitable for facilities with limited surface constraints, neighborhood proximity or requiring an extension. It is particularly adapted when compactness is essential, as it can integrate into the extension of a site with a limited space, as in small towns and tourist and resort areas. It is also suitable for mountain regions where, due to a more unstable climate, treatment facilities must be fully protected from bad weather. It is very tolerant to load variations. It can be used in secondary treatment for nitrification and denitrification.

Biostyr® Duo operation

Biostyr® Duo is a third-generation innovation in biofilter technology. It relies on the success of its Biostyr® predecessor – a compact solution allowing the treatment of carbon, chemical oxygen demand (COD), biochemical oxygen demand (BOD), nitrogen (NH_4-N and NO_3-N) and TSS – and is associated with a material tested in

MBBRs, the K5 support developed by AnoxKaldnes™, a subsidiary of Veolia Water Technologies. K5 supports are added to biostyrene beads and include polystyrene beads specifically manufactured and calibrated depending on the application. These beads are retained under the concrete ceiling of the structure (the cell), which is lined with nozzles allowing the removal of treated water. The aeration ramp is located at the bottom of the structure, which optimizes the time of contact between wastewaters and process air.

The biomass and the suspended solids accumulated in the media of the physical filter must be regularly removed by periodic backwashes. Due to the lightness of the material, the system can be washed by simple gravity expansion of the filtering media using the treated water stored in the upper part of the structure (water scouring).

The main characteristics of this process are as follows:

– Upward flow operation: wastewaters get into the reactor through the bottom and flow upwards, encountering various media layers.

– Biological treatment: when wastewaters cross the biofilm supports, attached bacteria convert ammonia into nitrates (nitrification), then nitrates into gaseous nitrogen (denitrification), thus removing nitrogen from the water.

– Filtration: filtration media remove suspended solids, guaranteeing that the effluent meets the required quality standards.

– Backwash: the reactor is subjected to regular backwashing to remove accumulated solids and maintain optimal performances.

Advantages of Biostyr® Duo

– High efficiency: the design of UMMR provides a large surface for the biological activity, which leads to efficient removal of nitrogen.

– Compact footprint: due to its integrated nature, the process minimizes the total required space, which makes it fit for various contexts.

– Reduced energy consumption: the upwards flow design minimizes the pumping and aeration needs, contributing to a reduction in energy costs.

– Low maintenance: the system is designed for easy use and minimal maintenance.

– Flexibility: the process can be customized to meet specific wastewater characteristics and treatment objectives.

Given the market competition, new economic stakes boost future innovation. Various research directions are suggested:

– higher flexibility for the primary treatment, in order to minimize reagent consumption;

– higher SS load;

– higher COD and nitrogen load;

– higher treatment rate as soon as the quality of the water to be treated allows it.

Given its higher density compared to polystyrene beads, this material is naturally positioned under biostyrene.

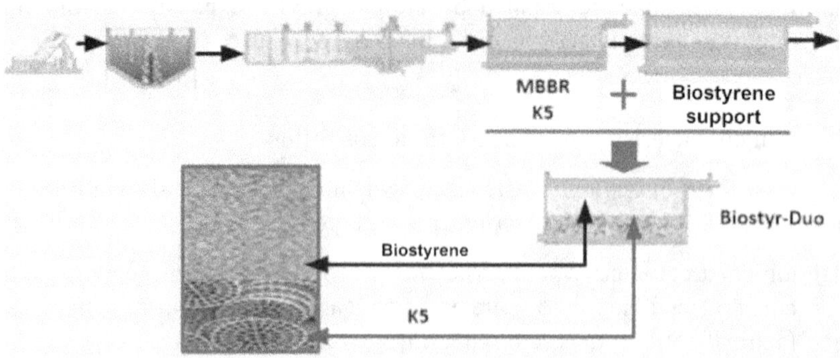

Figure 3.59. *Biostyr® and MBBR replaced by Biostyr® Duo*

Combining biological and physical treatment methods, this technology offers solutions for the removal of nitrogen and suspended solids, yielding cleaner and safer water. This approach is in line with the growing demand for environmentally friendly and cost-effective wastewater treatment solutions.

The use of Biostyr® Duo in pre-denitrification with a K5 height of 1.2 m and a height for the biostyrene beads of 2.5 m (granulometry of 4.5 mm) allows a more significant load reduction than a classical Biostyr® (3.5 m of beads) in the same configuration (1.2 kg.m^{-3} total material.d^{-1} compared to 0.8 kg N.m^{-3}.d^{-1}). The duration of the filtration cycle is on average 25 hours longer.

Characteristics

– Compact process requiring no additional clarification or filtration.

– Small footprint.

– Modular design, hence a rapid response to high load variations.

– Adaptability to various applications: carbon removal only; carbon removal and ammonia oxidation (nitrification); carbon removal associated with pre-denitrification, nitrification in distinct structures; carbon removal and simultaneous nitrification–denitrification; and tertiary nitrification or tertiary denitrification with the addition of a soluble carbon source, for example, methanol.

– Useless downstream clarifiers or filters, hence a reduction in operating and maintenance needs.

– Exceptional quality for treated water, even in very cold climates.

– No media leakage from the process and no need for reload or periodic renewal, which leads to a reduction in maintenance costs.

– No specific screening: classical screen rakes at the top of the system are sufficient.

– Additional nitrification and/or denitrification in a single process requiring minimal space.

– Fully automated process controlled by sensors, for reduced maintenance.

3.4. Conclusion

According to the manufacturers, this process has certain advantages, some of which are not specifically related to Biostyr® or Biofor® technology. In general terms, biofiltration has the following advantages.

First of all, it has the advantage of being a compact process, much more compact than activated sludge treatment, for example, and with a small footprint. Then, due to the principle of biofiltration, the process can do two in one, SS filtration and biological treatment. Nevertheless, the capacity to capture SS is limited by rapid clogging. Generally, biofiltration operates in modules of biofilters. The separation into several modules is in fact not an actual advantage but a necessity. Backwashing a biofilter requires turning off its treatment function. The separation into several modules allows washing a module and continuing biofiltration on the other modules, then alternating them.

Some of the disadvantages related to Biodrof® technology are:

– The odors and bioaerosols are due to the supply systems of Biodrof® biofilters, which, unlike Biofor® and Biostyr® technologies, are supplied with foul water at the top. To avoid odor diffusion, a choice was made to install these biofilters inside the building to contain odors. This involves additional costs.

– The sensitivity of biofilters to settleable SS. Indeed, biofilters are highly dependent on the performances of upstream primary settling tanks. The SS that could not be settled can rapidly lead to clogging of biofilters if they arrive in too high an amount. This explains the presence of primary settling tanks;

– The complexity of the process: as it involves both filtration and biological film notions, the process becomes difficult to model. All the more so as the air and substrate distribution in the filtering media may lead to the appearance of various sorts of microbial populations and various physico-chemical conditions affecting the media height.

3.5. References

Aesoy, A., Odegaard, H., Bach, K., Pujol, R., Hamon, M. (1998). Denitrification in a packed bed biofilm reactor (Biofor). Experiments with different carbon sources. *Water Res.*, 32(5), 1463–1470.

Agence de l'eau et ministère de l'environnement (1994). L'assainissement des agglomérations : techniques d'épuration actuelles et évolutions. *Cahier technique*, 27, 89–90.

Albertson, O.E. (1989). Optimizing rotary distributor speed for trickling filters. *Water Pollut. Control Federation Operators Forum*, 2(1), 57–75.

Atkinson, B. (1963). Recirculation, reaction kinetics, and effluent quality in a trickling filter flow model. *J. Wat. Pollution Control Fed.*, 35(10), 1307–1318.

Aubry, G. (2003). Enlèvement de l'azote des eaux usées par un procédé à culture fixée immergée. MPhil Thesis, Laval University, Quebec.

Aws, W.F. (1962). Transient operation of the trickling filter. *J. Sanit. Eng. Div. Am. Soc. Civ. Engrs*, 88(SA3), 21–29.

Bennett, S.H. (1963). Morphological aspects of extracellular polysaccharides. *J. Histochem. Cytochem.*, 11(1), 14–23.

Beyenal, H., Sani, R.K., Peyton, B., Dohnalkova, A.C., Amonette, J., Lewandowski, Z. (2004). Uranium immobilization by sulfate reducing biofilms. *Environ. Sci. Technol.*, 38(7), 2067–2074.

Boari, G., Mancini, I.M., Trulli, E. (1997). Technologies for water and wastewater treatment. *Aspects économiques de la gestion de l'eau dans le bassin méditerranéen*, 31, 262–287.

Boeglin, J.C. (1998). Traitements biologiques des eaux résiduaires. *Techniques de l'Ingénieur*, J3942 v1, 156–190.

Bruce, A.M. and Merkens, J.C. (1973). Further studies of partial treatment of sewage by high-rate biological filtration. *Wat. Pollut. Control*, 72, 499–523.

Canler, J.P., Durand, C., Perret, J.M. (1996). Efficacité des biofiltres vis-à-vis de l'azote. In *Actes de colloque sur le traitement de l'azote: cas des eaux résiduaires urbaines et des lisiers*. Lyon.

Daddles, G.A., Richardson, E.E., Barth, E. (1974). Plastic packing trickling filters for biological nitrogen control. *J. Water Pollut. Control Federation*, 46, 937–946.

Degrémont (2017). *Memento technique de l'eau*, 10th edition. Degrémont-Suez, Paris.

Deronzier, G., Schétrite, S., Racault, Y., Canler J.P., Héduit, A., Duchène, P. (2001). Traitement de l'azote dans les stations d'épuration biologique des petites collectivités. Technical Document, FNDAE [Online]. Available at: https://www.fndae.fr/documentation/PDF/fndae25.pdf.

Eckenfelder, W.W. (1963). Trickling filter design and performance, part 3. *Trans. Am. Soc. Civ. Engrs*, 128, 371–386.

Fucbu, Y. and Nakamura, M. (1988). Application of biological aerated filters to industrial wastewater treatment. In *61st Annual WPCF Conference*, Dallas.

Gilles, P. (1989). Industrial scale applications of fixed biomass on the Mediterranean coast: Design and operational results. In *IAWPRC Specialised Conference on Technical Advances in Biofilm Reactors*, Nice.

Gilles, P. (2004). Les Biofiltres, cours de l'eau OTV. Report, OTV-Veolia.

Gilles, P. and Bourdon, F. (1987). Nitrification and denitrification with fixed bacteria. *L'Eau, L'Industrie, Les Nuisances*, 93, 225–243.

Henze, M., Harremoes, P., La Cour Janssen, J., Arvin, E. (1997). *Wastewater Treatment – Biological and Chemical Processes*, 2nd edition. Springer-Verlag, Berlin.

Kantardjieff, A. (1989). Le premier filtre canadien pour le traitement des eaux usées du Canton d'Orford (Québec). *Sciences et techniques de l'eau*, 22(1), 73–82.

Kornegay, B.H. and Massey, M.L. (1971). A rational approach to trickling filter design. In *162nd Meeting Nat. Am. Chem. Society*, Washington, D.C.

Lamb, R. and Owens, S.G.H. (1970). A suggested formula for the process of biological filtration. *Wat. Poll. Control*, 69, 209–220.

Lee, K.M. and Stensel, H.D. (1986). Aeration and substrate utilization in a sparged packed-bed bio-film reactor. *Journal (Water Pollution Control Federation)*, 58(11), 1066.

Leglise, J.P., Gilles, P., Moreaud, H. (1980). A new development in biological aerated filters. In *53rd Annual WPCF Conference*, Las Vegas.

Lewandowski, Z. and Boltz, J. (2011). Biofilms in water and wastewater treatment. In *Treatise on Water Science*, Wilderer, P. (ed.). Elsevier, Amsterdam, 529–570.

Majer, W.J., Behn, V.C., Gates, S.C.D. (1967). Simulation of the trickling filter performance. *J. Sanit. Eng. Div. Am. Soc. Civ. Engrs*, 93(SA4), 471–489.

Metcalf & Eddy Inc. (2002). *Wastewater Engineering: Treatment and Reuse*, 4th edition. McGraw-Hill, New York.

Moreaud, H. and Gilles, P. (1979). Nitrogen removal from wastewater. *T.S.M. l'eau*, 74, 241.

Morgenroth, Y. (2008). Modelling biofilm systems. In *Biological Wastewater Treatment – Principles, Modelling, and Design*, Henze, M., van Loosdrecht, M.C.M., Ekama, G., Brdjanovic, D. (eds). IWA Publishing, London.

National Research Council (1946). Sewage treatment at military installations. Report of the subcommittee on sanitary engineering. *Sewage Work. J.*, 18, 787–1096.

Oleszkiewicz, J. (1975). Efficiency of plastic media trickling filters operating under extreme organic loadings. *Progr. Water Technol.*, 7(3/4), 607–615.

Oleszkiewicz, J. (1977). Theory and design of high-rate plastic media trickling filters. *Progr. Water Technol.*, 9, 777–785.

Oleszkiewicz, J. and Eckenfelder, W.W. (1974). The mechanism of substrate removal in high rate plastic media trickling filters. Technical Report 33, EWRE, Vanderbilt University Press, Nashville.

Paffoni, C. and Payraudeau, M. (1998). Le prototype Biostyr : bilan des essais. *Tech. Sci. Méth.*, 3, 19–28.

Paffoni, C., Gousailles, M., Rogalla, F., Gilles, P. (1989). Aerated biofilters for nitrification and effluent polishing. In *IAWPRC Specialised Conference on Upgrading of Wastewater Treatment Plants*, Munich.

Payraudeau, M. and Bontonou, J.Y. (1989). Traitement biologique à basse température selon un procédé de filtration biologique aérée : le Biocarbone. In *69ème congrès de l'AGHTM*, Quebec.

Payraudeau, M., Banquet, G., Rogalla, F., Bourbigot, M.M., Sibony, J., Gilles, P. (1988). Nutrient removal with biological aerated filters. In *61st WPCF Conference*, Dallas.

Payraudeau, M., Sauvegrain, P., Pénillard, P. (1989). Nitrification secondaire et tertiaire des eaux froides par le procédé à biomasse fixée Biocarbone. In *12ème symposium international sur le traitement des eaux usées*, Montreal.

Payraudeau, M., Paffoni, C., Gousailles, M. (2000). Tertiary nitrification in an upflow biofilter on floating media: Influence of temperature and COD load. *Water Sci. Technol.*, 41(4), 21–27.

Pujol, R., Lemmel, H., Vedry, B. (1997). Le Biofor en nitrification tertiaire : le prototype SIAAP de la station Seine Aval (Achères). In *Actes de colloque sur la nitrification tertiaire par cultures fixées*, Paris.

Rijn, J., Tal, Y., Schreier, H.J. (2006). Denitrification in recirculating systems: Theory and applications. *Aquacult. Eng.*, 34(3), 364–376.

Rincke, G. and Wolters, N. (1970). Technology of plastic medium trickling filters. In *Proc. 5th Int. Conf. Wat. Pollut. Res.*, San Francisco.

Rocher, V., Paffoni, C., Goncalves, A., Legaigneur, V., Dutot, A., Gousailles, M. (2006). Le traitement des eaux usées par biofiltration : cas de la station Seine Centre (SIAAP). *Techniques sciences et méthodes*, 10, 105–125.

Rocher, V., Paffoni, C., Gonçalves, A., Azimi, S., Gousailles, M. (2008). La biofiltration des eaux résiduaires urbaines : retour d'expérience du SIAAP. *Revue des sciences de l'eau*, 21(4), 475–485.

Rogalla, F. and Goncalves, R.F. (1990). Temperature influence on complete nitrogen removal with B.4F biological aerated filters. In *25th IAWPRC Conference*, Kyoto.

Sibony, J. (1984). Industrial applications of fixed biomass for wastewater treatment. *L'Eau, L'Industrie, Les Nuisances*, 84, 31–42.

Sonthonnax, A. (2006). Influence du substrat carboné sur les cinétiques de dénitrification des eaux usées. Thesis, Paris-East Créteil University, Paris.

Stensel, H.D., Brenner, R.C., Lee, K.M., Melcer, H., Rakness, K. (1989). Biological aerated filter evaluation. *J. Env. Eng.*, 6(3), 655–671.

United States Environmental Protection Agency (2000). Wastewater technology fact sheet trickling filter. Office of Water, Washington, D.C. [Online]. Available at: https://www3.epa.gov/npdes/pubs/trickling_filter.pdf.

Water Environmental Federation (WEF) (2000). *Aerobic Fixed-growth Reactors*. Water Environment Federation, Alexandria.

Wilkinson, J.F. (1958). The extracellular polysaccharides of bacteria. *Bacteriol. Rev.*, 22, 46–73.

Zobell, C.E. (1943). The effect of solid surfaces upon bacterial activity. *J. Bacteriol.*, 46(1), 39–56.

4

Rotatory Biological Contactors

4.1. General information and rotatory biological contactor ecology

Rotatory biological contactors (RBCs) are an aerobic biological treatment system with fixed biomass. This process can be considered as one of the aerobic biological treatment systems in which the bacteria culture is fixed on a support, as is the case with trickling filters, biofilters or hybrid systems.

The supports of purifying microflora are RBCs partially immersed in the effluent to be treated and subjected to a rotational motion leading to both mixing and aeration. During the immersed phase, the biomass absorbs the organic material it degrades by aerobic fermentation thanks to the atmospheric oxygen. As soon as it exceeds a thickness of several millimeters, the excess biofilm (sludge) detaches and is driven to a liquid–solid separation stage. The sludge thus trapped is automatically sent by periodic pumping to the head structure to be stored and digested.

Figure 4.1. *Open rotatory biological contactors (Veolia)*

Biofilm plays an essential role in the performance of RBCs for wastewater treatment. RBCs provide a surface for the growth and fixation of microorganisms that form a biofilm. This biofilm is a complex community of microorganisms, particularly bacteria, fungi and protozoa, which work together to decompose and remove the organic material and other contaminants of wastewaters. The microfauna of RBCs may differ depending on whether they are located in the middle, at the center or on the edge of the RBCs. Protozoa are often encountered in all compartments. They can develop in the zooglea, *Plateyma*, *Aspidisca*, *Oxytrichia*, *Uronema*, etc. On the edge of RBCs, there are colonies of protozoa next to diatoms such as *Nitzschia* and *Navicula*.

The principle of the RBC-based treatment process is identical to that of biofilters or that of activated sludge; namely, degradation of the pollution by the bacterial mass along two large paths: catabolism and anabolism.

In this process, oxygen becomes an important parameter for the degradation of organic materials.

4.2. Implementation of RBCs and wastewater quality

Although apparently simple, the treatment process by RBCs involves complex processes. The degradation of the substrate by the biomass fixed on the disks is limited by the diffusion of the substrate through the biofilm, by the oxygen transfer, by the biochemical reactions themselves and by the physical characteristics of the process.

These physical characteristics have a direct influence on the conditions of contact between substrate and microorganisms, on the contribution of oxygen and on maintaining a biofilm of more optimal thickness.

The RBCs consist of several thin polypropylene disks mounted on a horizontal axis to form a drum. This axis is fixed above a tank, in which passes the liquid to be treated. Driven by an engine, this axis induces the rotation of disks. Each of the implemented drums has 10 to 200 disks.

The microorganisms responsible for the degradation are naturally fixed on the disks and form a biofilm of a thickness that may reach 5 mm. About 40% of the surface of the disks is immersed. The rotational motion of the disks about the axis exposes the biomass alternatively to the atmosphere and to wastewaters, thus allowing aeration and mixture.

Figure 4.2. *Groups of RBCs (Veolia)*

Shearing forces created by the rotational motion limit the thickness of the biofilm and drive a detachment of excess biomass, which is then separated from the effluent by means of a secondary clarification technology.

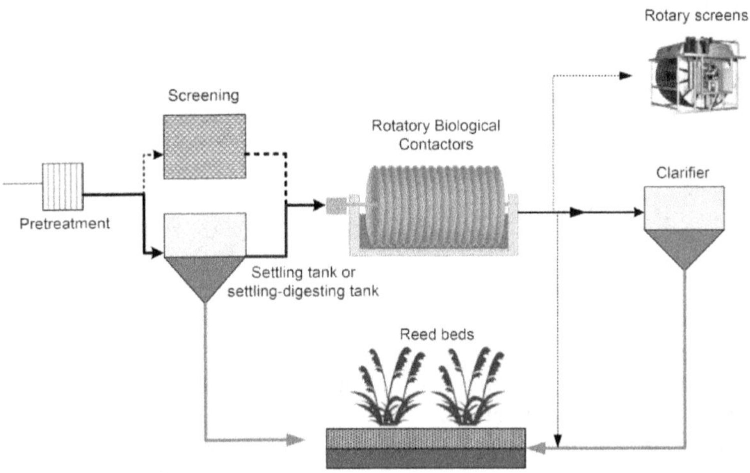

Figure 4.3. *Rotatory biological contactor treatment system*

The disks stacked side by side are threaded rods bolted on metal profiles, which in their turn are fixed on elements secured to the shaft. The set forms a robust and undeformable drum. The spacing between disks is about 2 cm. The disks plunge, by half of their diameter, into semi-cylindrical tanks. The disks slowly turn about a horizontal axis so that the bacteria culture present on the support is alternatively in contact with water and air. The direction of rotation has a great importance in terms of purification, as the disks must rotate in the direction of water circulation, and the velocity of water circulation in contact with bacteria is higher.

The rotational velocity of disks is also an important factor, as it must:

– not deprive the culture of oxygen through too long an immersion;

– allow homogeneous stirring of the liquid to facilitate exchanges between the liquid and the bacteria mass;

– avoid an ejection of the bacteria mass from too high rotational velocities by adopting a velocity of about 20 m.min^{-1} or on average 2 rpm.

The immersion phase thus allows bacteria to degrade the present organic materials. Rising to the surface phase allows them to take from the atmosphere the oxygen required by the respiratory needs of microorganisms.

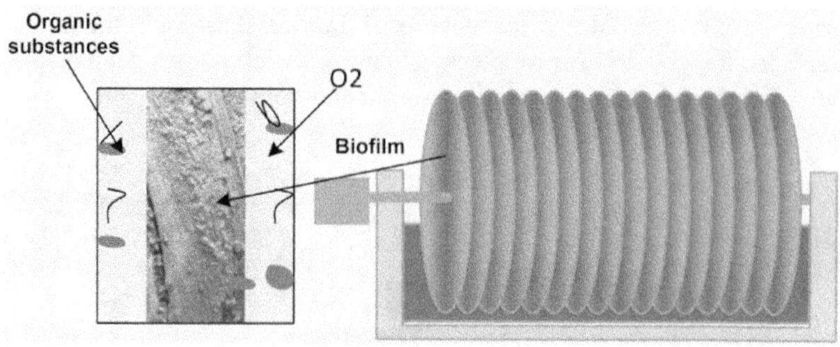

Figure 4.4. *Simplified diagram of the RBC process*

Without prior seeding, bacteria and their immediate predators get fixed on the disks and form the zooglea, also known as biological turf or fixed biomass. Bacteria fixation results from a combination of various forces: electrostatic attraction, hydrophobic interaction, Van der Waals forces and Brownian motion, thus producing a permanent or temporary attachment. Disk rotation allows alternating exposures of the biomass to the air and residual water. Microorganisms will then be able to draw their nutrients (organic materials, various nitrogen and phosphorus forms) from the effluent and oxygen from the air. Depending on thickness and therefore on the ability to get oxygen by diffusion in its various strata, the biofilm is the location of complex biochemical reactions. The approximate thickness of the biofilm is highly variable. It ranges between 250 and 3,000 µm. When the biomass layer is too significant, shearing forces become predominant and lead to the detachment of one part of the biofilm. Treated waters then flow into a lamellar settling tank, in which suspended solids are filtered. Sludge is pumped at the bottom of the tank to be recirculated with floating agents to the all-water tank, where it is stored.

In fact, an RBC treatment plant should involve the following structures:

– pretreatment (screening–grit removal–oil removal);

– primary settling tank, to avoid the accumulation of coarse and heavy materials at the bottom of the tank or a screening;

– rBCs;

– liquid–solid separation: clarifier, rotary screens, flotation or extensive process (reed beds, lagoons).

The technical choice of treatment upstream of disks depends on several criteria, such as the quality of the effluent to treat, purification performances to be reached by the plant, constraints in terms of sludge treatment and the technology separating the biomass downstream of disks.

Special attention is given to the primary treatment in the case of strict effluent standards.

The quality of treated water is directly related to the pollutant load applied per unit time and the wet surface of the disks. For effluent waters, the expected quality is BOD_5: 25 to 35 $mg.L^{-1}$, COD: 90 to 125 $mg.L^{-1}$, SS: 30 $mg.L^{-1}$, TN: 50 $mgN.L^{-1}$, TKN: 15 $mgN.L^{-1}$ (if specific dimensioning) and total P: average efficiency of about 25% (additional treatment by iron salt injection considered).

4.3. Dimensioning of RBC facilities

The dimensioning of RBCs varies depending on the raw effluent quality, the load to be treated and the expected effluent levels.

4.3.1. *Applied surface load*

The applied load has an essential role in the performance of RBCs. It determines the quantity of organic material applied per unit time and per m^2 of surface. RBC performance strongly depends on the capacity of microorganisms of the biofilm to degrade the organic material. If the load is too high, microorganisms might be unable to consume the entire BOD_5, which leads to an incomplete treatment and a low removal efficiency. It is consequently important to choose an adapted load depending on the characteristics of wastewaters, the design of RBCs and the expected performances. The optimal load of RBCs may vary depending on factors such as the concentration and biodegradability of the organic material, temperature, pH and residence time. Overloading the system reduces the treatment efficiency, while a too low load underuses the available treatment capacity. The retained surface load depends on the expected outlet effluent temperature and quality. Table 4.1 summarizes the quality of effluents depending on the applied loads (g $BOD_5.m^{-2}.d^{-1}$).

Organic loading rates (g $BOD_5.m^{-2}.d^{-1}$)	BOD_5 (%)	COD (%)	TN
4–6	< 30	< 90	TN < 20 $mg.L^{-1}$ 90% NH_4
6–8	< 30	< 90	80% NH_4
8–10	30	120	70% NH_4
10–12	30	120	< 10% NH_4
12–14	< 60	< 60	< 10% NH_4

Table 4.1. *Performances of RBCs depending on the applied BOD_5 load (at 15°C)*

4.3.1.1. *Total surface*

The total surface of disks depends on the inlet flow rates and BOD_5 concentration of raw wastewater. We have:

$$\text{Outlet } BOD_5/\text{inlet } BOD_5 = (1 - ab)\ 10^{-F(Sc)} + ab\ 10^{-F'(Sc)}$$

$$F(Sc) = -Xb_1(\frac{Sc}{Q} - 0.8)$$

$$F'(Sc) = -Xb_2(\frac{Sc}{Q} - 0.8)$$

In this relation, the terms a, b, Xb_1 and Xb_2 are given by the following relations:

$$ab = 0.8\ BOD_{inlet}^{-0.375}$$

$$Xb_1 = 0.0456\ BOD_{inlet}^{-1.78}$$

$$Xb_2 = 0.0538\ BOD_{inlet}^{-0.89}$$

Moreover, the influence of temperature is highlighted here in a special manner, so that:

$$f(T){:}\ 10^{0,02(T)} - 1.713\ 10^{-0.1\,(T)} - 3.5\ 10^{0.1\,(T)}$$

The total surface of disks to be implemented becomes:

$$St = f\left(\frac{r}{D}\right) f(R) f\left(\frac{1}{T}\right) f(Q) BOD\ inlet$$

where:

– r is the distance from the disk drum shaft to the liquid level;

– D is the diameter of disks;

– f (R) = $(1 - R)^{0.25}/R^{0.4}$, R being the treatment efficiency;

– T is the temperature (°C);

– Q is the ratio of influence of flow rates on the surface of disks.

Another empirical relation proposes:

$$Sc = \frac{QLo}{Csurf}\frac{(Lo - Lf)^{1.4}}{A\,Lf}$$

The contact surface is given in m^2 and represents the cultivated surface, Csurf is the applied surface loading (kg $BOD_5.m^{-2}.d^{-1}$), and A is a dimensionless factor that depends on load and temperature and may vary between 145 and 400.

4.3.1.2. *Applied organic load*

Calculation of the global BOD_5 load received by the RBCs:

– hypothesis: 1 Eqh = 60 g $BOD_5.d^{-1}$;

– C_{BOD5}: organic load received by the disks in kg $BOD_5.d^{-1}$;

– C_{EH}: grams of BOD_5 per Eqh and per day;

– R pretreatment: % BOD_5 reduction due to pretreatment;

– N_{EH}: number of connected inhabitants;

– $C_{BOD5} = N_{EH} \times C_{Eqh} \times 10^{-3} \times (1 - (R/100))$.

EXAMPLE.–

– Plant of 800 Eqh equipped with a settling tank.

– BOD_5 required at the outlet: 30 $mg.L^{-1}$.

– Nitrogen removal: TN = 20 $mg.L^{-1}$.

– $C_{BOD5} = 800 \times 60 \times 10^{-3} \times (1 - [10/100]) = 43.2$ kg $BOD_5.d^{-1}$.

– For an external temperature of 12°C, a surface load of 6 g of $BOD_5.m^{-2}.d^{-1}$ is considered.

– The required surface of disks is: S = 43.2 x 1,000/6 = 7,200 m^2.

4.3.2. *Geometry of RBCs*

RBCs generally have a diameter equal to 2 or 3 m, with a thickness ranging between 7 and 14 mm. The drum has a maximum of 200 disks to allow the shaft to support the entire weight, in addition to the weight of the biological turf, which reaches nearly 6 kg/m^2 for the two faces. A motor reducer assembly ensures the rotation of the group of batteries. The disks are immersed close to the shaft to make the most of the available surface.

The space between the disks and the bottom or the edges of these basins is only several centimeters to avoid creating dead zones at the bottom of the tanks and especially to prevent the accumulation of suspended solids. The disks are generally sheltered in a closed and covered location in order to take advantage of good climate conditions when they exist.

These locations are aerated by steerable shutters. Energy consumption depends essentially on the disk diameter and number. Energy consumption is summarized in Tables 4.2 (2 m diameter disk) and 4.3 (3 m diameter disk).

Number of disks	30	40	50	60	70	90	120	160	180	200
Installed capacity (Cv)	0.25	0.25	0.25	0.25	0.25	0.25	0.5	0.5	0.5	0.5
Absorbed power (kW)	0.045	0.06	0.075	0.09	0.10	0.13	0.18	0.24	0.27	0.3

Table 4.2. *Characteristics of the facility for 2 m diameters*

Number of disks	30	40	50	60	70	90	120	160	180	200
Installed capacity (Cv)	0.25	0.25	0.25	0.25	0.5	0.5	1.0	1.0	1.0	1.0
Absorbed power (kW)	0.075	0.10	0.12	0.15	0.17	0.20	0.30	0.40	0.45	0.50

Table 4.3. *Characteristics of the facility for 3 m diameters*

The higher the number of stages, the better the performances, as a more significant biomass is attached to the disks. Moreover, the last stages facilitate the process of effluent nitrification.

The diameter of disks ranges between 1.2 and 4.5 m for operational plants, but it is most often between 2 and 3 m. The transfer of oxygen varies according to the disk diameter and the rotational velocity. The lower the diameter, the higher the efficiency (Wilson et al. 1980).

4.3.3. *Hydraulic loading rates*

The process can support wide variations of hydraulic and organic loading rates, without significant loss of biomass.

4.3.4. *Residence time*

The residence time of the effluent in the tank must be long enough to allow good degradation of the organic material and also short enough to avoid overloading the first RBCs to the extent that it is a plug flow that crosses the structure. A residence time ranging between 1 and 2 hours is generally recommended. When the residence time increases, the size of the SS increases. This parameter expresses the time needed for the effluent to cross the bioreactor, which may influence the time of retention of microorganisms and the efficiency of pollutant removal. It is therefore a dimensioning parameter. An optimized residence time allows the degradation of pollutants by microorganisms, which is reflected by high removal efficiency. On the one hand, too long a residence time may lead to the accumulation of inert materials and to the growth of less efficient microorganisms, thus reducing the global performance of the system. On the other hand, if this parameter is too short, this reduces the time available for microorganisms to degrade pollutants, which is reflected by lower removal efficiencies. The optimal contact time varies as a function of the characteristics of wastewaters, process design and expected performances.

4.3.5. *Liquid–solid separation*

The SS concentration of the effluent of the basin of disks is much lower than that of activated sludge (about 200 mg SS.L^{-1}). Moreover, the settling of fine particles is not facilitated by flocculation and sludge layer conditions as in activated sludge.

The operation of the standard clarifier and its performances rely on meeting the following parameters:

– rising velocity: Vasc < 0.4 m.h^{-1} for an SS outlet < 30 mg.L^{-1};

– residence time, ranging between 1 and 2 hours.

These parameters are calculated for the peak flow (Qp).

If a lamellar clarifier is installed, it is dimensioned for the peak flow rate (Qp), a residence time ranging between 1.5 and 2 hours and a rising velocity (Vasc) below 0.8 m.h^{-1}.

The settling volume must meet the following criteria:

$$Qp \times 1.5 < \text{volume} < Qp \times 2$$

The settling surface is deduced from the rising velocity and the peak flow rate.

The available settling surface depends on the implemented settling cone:

$$S\ (m^2) = (Qp/Vasc)$$

The maximal rising velocity recommended induces a minimal residence time on the structure. An excessive residence time may lead to anaerobic reactions in the lamellar settling tank susceptible to intensifying the appearance of floats.

It is therefore important to avoid overdimensioning of the lamellar clarifier to limit the savage denitrification phenomenon, which is susceptible to facilitating the appearance of floats.

The clarifier can be replaced by a polishing pond (as the settling-digesting tank by a settling pond) and more recently by reed beds. In this last configuration, there is no settling-digesting tank and reed beds ensure both the separation between sludge and purified water and dehydration and sludge storage.

4.3.6. *Sludge production*

The quantity and quality will depend on the input effluent. Removal of 1–1.5 $kg.kg^{-1}$ of BOD_5 is given for an efficiency of 80% and of 0.8 $kg.kg^{-1}$ for an efficiency of 90% or more (Boutin 1976). Sludge production is related to the mass surface loading applied to the RBCs.

Sludge production can be considered near 40 g SS/d/Eqh in the case of a high nitrification objective (load ranging between 4 and 6 g $BOD_5.m^{-2}.d^{-1}$) and can reach over 50 g SS/d/EH in the case of a low purifying objective that corresponds to a load above 9 g $BOD_5.m^{-2}.d^{-1}$. These values can be expressed in kg SS per kg removed BOD_5, and we obtain values of approximately 0.8 kg SS per kg removed BOD_5.

Retained sludge productions are consistent with those of bacteria beds (0.8 kg SS/kg removed BOD_5) and activated sludge (1 kg $SS.kg^{-1}$ removed BOD_5 for the treatment of carbon or 50 g/EH/d and 0.8 kg $SS.kg^{-1}$ removed BOD_5 in the case of the treatment of nitrogen or 40 g/EH/d).

4.3.7. *Sludge retention time*

This parameter refers to the residence time of microorganisms in the bioreactor, affecting the concentration and the activity of biomass. A retention time (T_R) may range between 5 and 20 days, depending on the expected treatment levels. It can affect the performances of RBCs by influencing the composition and the activity of microorganisms in the biofilm. A high value for T_R facilitates the growth of microorganisms with slower growth, such as nitrifying agents, which are responsible for the oxidation of ammonia and nitrites. It contributes to the development of more stable microbial communities resistant to perturbations, which is reflected by more consistent and reliable treatment performances.

In contrast, too low a value for T_R facilitates the growth of microorganisms with faster growth, such as heterotrophic bacteria, responsible for removing organic material. Consequently, T_R must be well dimensioned to balance the efficiency and stability of the biomass. The increase in T_R beyond 15 days has both positive and negative effects on the biological performance and efficiency of the system. On the positive side, an increase in T_R drives a higher concentration of biomass, improving the efficiency of organic material removal and the global performance of the system treatment. The longer it is, the more the biomass grows and develops, producing a more established and more efficient biofilm on the surfaces of disks. However, the increase in T_R beyond 15 days also has negative effects. One of the main drawbacks is the increase in operational and maintenance costs associated with a higher biomass concentration in the reactor. Higher biomass concentrations require frequent maintenance, such as cleaning disks, to avoid clogging and reduce hydraulic resistance. Another negative effect is the risk of reduced treatment efficiency due to the accumulation of inert or inactive biomass, reducing the surface available for the microbial growth by hindering the nutrient mass and oxygen transfer to the active biomass.

4.3.8. *Energy consumption*

For a mechanical drive, the Environmental Protection Agency (EPA) gives 2.03 kWh/unit independently of the type of support (Brenner et al. 1984). For a rotational velocity of 1.6 t/min, 2.3 kWh per shaft for a standard support are consumed and 3.4 kWh for a high-density support.

4.3.9. *Removal of organic substances*

The efficiency of removing organic pollution depends on various factors, particularly on the characteristics of raw wastewaters, design and use parameters and the active biomass present on supports. Wastewater characteristics, such as the concentration and nature of the organic substances, influence the efficiency of BOD removal. The nature of the organic pollution, biodegradability and toxicity also have an impact on the efficiency of BOD_5 removal. The design and operating parameters of the bioreactor, such as the disk rotational velocity, the residence time and the temperature, influence the efficiency of BOD_5 removal. The rotational velocity of the disk determines the thickness of the biofilm and the time of exposure to wastewaters, thus affecting the efficiency of the treatment. The temperature influences the microbial activity and the growth kinetics of microorganisms. A high temperature facilitates the biological activity and involves an increase in efficiency. Overall, bioreactors reach high efficiencies in BOD_5 removal, with reported values ranging between 70% and 95%. An empirical relation was proposed by the EPA, related to the BOD_5 removal (Brenner 1984):

$$Cs/Ce = \text{Exp}\ (K\ (0.000125\ V/Q)^{0.5})$$

where Cs and Ce are respectively the output and input BOD (mg.L^{-1}), K is a constant of the reaction rate (0.3 to 13°C), V is the medium volume (m^3) and Q is the wastewater flow rate (L.s^{-1}).

4.3.10. *Nitrogen treatment*

Nitrification is a two-stage process: in the first stage, autotrophic ammonia-oxidizing bacteria (AOB) convert ammonium into nitrite, and the second stage is the conversion of nitrite into nitrate by nitrite-oxidizing bacteria (NOB). Heterotrophic bacteria develop more easily in high values of BOD than the autotrophic ones. Figure 4.5 shows the complete cycle of nitrogen, indicating the main oxic and anoxic processes. The competition between autotrophic and heterotrophic bacteria affects nitrification. Nitrification is the biological oxidation of the ammonia into nitrite and nitrate, while denitrification is the biological reduction of nitrate into gaseous nitrogen. The two processes require specific environmental conditions, particularly temperature, pH, OD and a precise contact time. A concentration of dissolved oxygen (DO) above 2 mg.L^{-1} facilitates the aerobic nitrification and below 0.5 mg.L^{-1} facilitates denitrification. The pH must be maintained between 6.5 and 8.5. Temperature is another essential factor affecting the removal of total nitrogen (TN), with optimal performances between 20 and 30°C. Overall, RBCs have shown excellent efficiency in removing TN, with rates ranging between 50% and 90%, according to the design parameters and the operating conditions.

The removal of ammonia nitrogen depends on several factors, such as the concentration of ammonia nitrogen in the raw effluent, the aeration rate, the temperature and pH. The mechanism for removing ammonia nitrogen in the RBCs involves the conversion of ammonia nitrogen into nitrite and then into nitrate by AOB and NOB. Nitrate is converted into gaseous nitrogen by denitrifying bacteria.

Nitrification can occur in the same unit of RBCs following the removal of carbon material if the nitrifying biomass still has a sufficient support surface. It can also be considered in specifically designed units, when the plant is equipped with several serial stages. Most bacteria involved in nitrification (Nitrosomonas, Nitrobacter) are known as autotrophic, which means they draw their energy from the reduction of carbon dioxide and bicarbonates.

The aerobic treatment of pollution by fixed cultures yields excellent nitrification efficiencies by adapting the hydraulic and organic loading rates used with RBCs to the demands of nitrogen treatment.

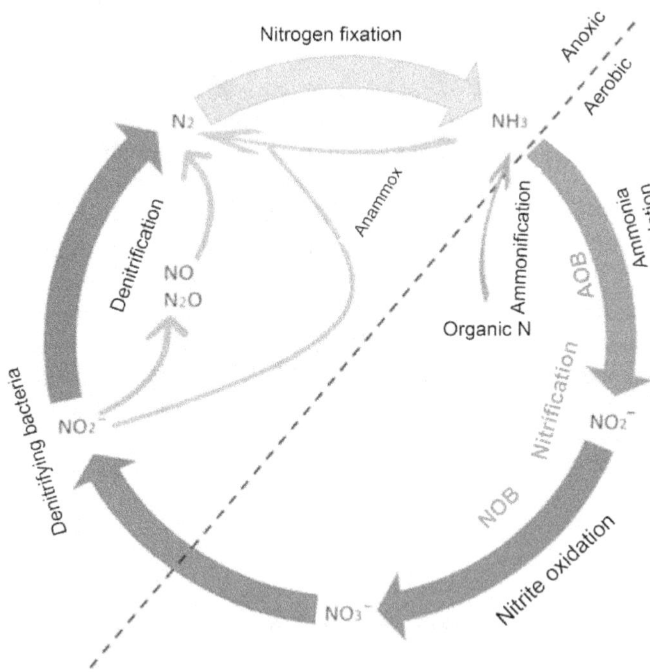

Figure 4.5. *Nitrogen cycle summarizing the main processes, aerobic and anoxic*

Nitrification is very sensitive to variations of organic loading rates; this is why the removal of organic substances must have occurred in the first stage or the two first stages. A maximum organic loading rate below 5 g total $BOD_5.m^{-2}.d^{-1}$ is needed. Nitrification can unfold at the end of a row of disks, when the organic loading rate is not too high, as well as in a stage separated from the one involving treatment of the organic material. The two phases of the process can be identified based on the biomass: the bacteria film is darker and finer when there is nitrification.

When nitrification is expected after the treatment of organic substances, dimensioning is conducted stage by stage, each of them being considered a single reactor. Similar to organic substance degradation, nitrification depends on rotational velocity (Boiler et al. 1990), in relation to the increase in available DO.

RBCs reach high efficiencies for removal of ammonia nitrogen, from 70% to 95%, depending on the inlet concentration of ammonia nitrogen and on the operating conditions. The optimal pH range for the removal of ammonia nitrogen is between 7.5 and 8.5. Low temperatures inhibit the growth and activity of nitrifying bacteria, which leads to a decrease in ammonia nitrogen removal efficiency. High ammonium concentrations generate incomplete nitrification and an accumulation of nitrites, which are toxic for Nitrosomonas microorganisms.

The biofilm of a bioreactor includes several layers, the more external ones containing aerobic bacteria that degrade the organic material and produce carbon dioxide and water. The deeper layers of the biofilm may contain anaerobic bacteria that decompose more complex organic compounds. The biofilm is constantly exposed to the oxygen during the rotation of disks, which facilitates the growth and activity of aerobic microorganisms. The deeper layers of the biofilm can be anoxic or anaerobic, which means they are not exposed to oxygen. Microorganisms can use various electron acceptors to decompose the organic material in these layers. In the anoxic layer, microorganisms can use nitrate or other oxidized nitrogen compounds such as electron acceptors, while in the anaerobic layer, they can use sulfate or other oxidized sulfur compounds.

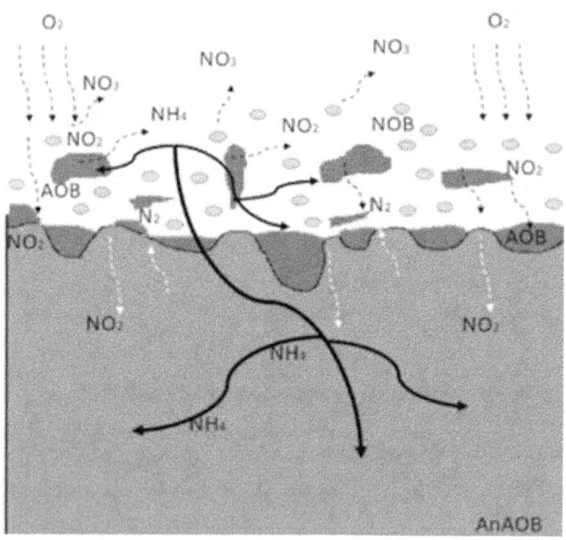

Figure 4.6. *Distribution of various types of microorganisms in the biofilm*

The presence of anoxic and anaerobic layers in the biofilm allows the complete removal of organic substances and nutrients from wastewaters. The biofilm acts as a physical barrier that filters and removes the contaminants while providing an environment favorable to the growth and activity of microorganisms. Overall, the structure and function of the biofilm are complex and dynamic, with various layers of microorganisms fulfilling various functions in the removal of contaminants from wastewaters. The presence of aerobic, anoxic and anaerobic layers contributes to more efficient and complete removal of organic substances and nutrients from wastewaters, which makes RBCs an efficient option for wastewater treatment.

Recirculation allows the dilution of inflowing organic substances, when not too significant, and facilitates nitrification.

Denitrification is due to heterotrophic bacteria in an anoxic environment. These bacteria transform nitrates (NO_3^-) into atmospheric nitrogen N_2. Denitrification can occur in two ways: either inside the deepest layers of the biofilm or in a completely separated treatment stage, after nitrification.

In the deep layers of the biofilm, denitrifying and nitrifying bacteria coexist in the biofilm, with denitrification taking place only when oxygen transfer has a sufficiently low efficiency to allow the formation of an anoxic environment. Nitrogen removal is influenced by the oxygen partial pressure in the air of the above-water part, water temperature, hydraulic residence time and C/N ratio. To obtain maximum removal using aerobic disks, recirculation (100%) issued from the tank of RBCs must be implemented towards the denitrification compartment located at the beginning of the process. This anoxic tank has the classical technical characteristics allowing the development of denitrifying flora. The global residence time calculated for two storage hours at the peak flow rate must allow the maintenance of anoxic conditions favorable to denitrification ($O_2 < 0.5$ mg.L^{-1}).

The recirculated effluent would bring the nitrates and the non-recirculated part would bring the carbon. By adapting an anoxic zone, it is possible to obtain good denitrification results (outlet[TN] < 20 mg/l).

4.3.10.1. *Phosphate removal*

Physico-chemical phosphate removal can be associated with the process after $FeCl_3$ injection, leading to outlet values (PT) < 2 mg.L^{-1} or 80% reduction.

$FeCl_3$ injection is proportional to the inlet flow rate (potential automatic regulation by an electromagnetic flow meter at the inlet). It is recommended to propose at least two distinct injection points.

4.4. Factors influencing the RBC process

A certain number of physical or biological parameters influence the performances of RBCs.

4.4.1. *Choice of the liquid–solid separation technique*

Compared to a clarifier, the filtering drum offers better performances in pollution reduction and better stability for the effluent quality. In the case of restrictive effluent standards, it is often preferred to conventional or lamellar clarifiers. The filtering drum can also be used for a plant whose treatment objectives are the removal of carbon pollution and nitrification. Indeed, nitrification leads to nitrate concentrations in the treated effluent above 10 $mg.L^{-1}$, which is incompatible with the implementation of a clarifier due to wild denitrification at the bottom of the structure. Indeed, the formed nitrates will be partly removed by endogenous denitrification in the clarifier, the consequence being a rise of sludge to the surface and a degradation of the quality of the treated effluent, risking non-compliance.

4.4.2. *System hydraulics*

Hydraulics play an important role both in terms of nutrient input for the bacteria and in terms of implantation of microorganisms. Growth scattered on the disks may be a consequence of poor hydraulic design. In RBCs, there is a plug flow. However, even in the case of a plug flow between stages, there is a complete mix inside each compartment.

4.4.3. *Influence of DO*

The quantity of oxygen dissolved in the liquid to be treated is sometimes used as an explanation for a poor efficiency: in too low quantity, it would become the limiting factor of the reaction. A low O_2 concentration would allow the development of sulfate-reducing bacteria, such as Beggiatoa. A minimum rate of 1 to 3 mg of $O_2.L^{-1}$ is recommended.

Oxygen diffusion depends essentially on its concentration at the surface. As oxygen diffuses much more rapidly in the air than in water, this influences the oxygenation of biological turf. Figure 4.7 shows the velocity of O_2 diffusion in the biological turf.

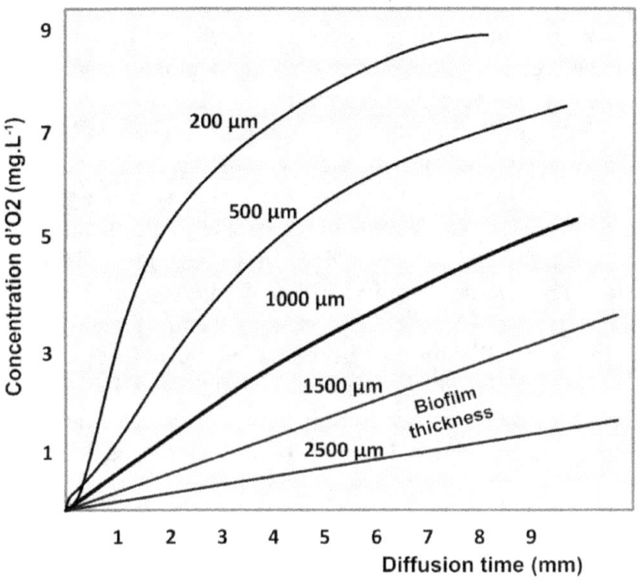

Figure 4.7. *Oxygen diffusion in the biofilm*

For various thicknesses of the biofilm, rapid diffusion of oxygen can be observed. For example, 3 minutes are needed for the O_2 to penetrate a layer of 200 μm thickness, as O_2 depends on its value at the air–water interface. Tests have shown that this layer thickness is maximum and suggest the existence of an anaerobic zone at the interface with the support.

4.4.4. *Rotational velocity*

The rotational velocity of disks influences the shear stress on the biofilm and the transfer of oxygen and nutrient mass to microorganisms. The optimal rotational velocity depends on the specific application of wastewater treatment, ranging between 1 and 10 turns per minute. It is an essential parameter in the design and operation of RBCs. The rotational velocity of the disk impacts the thickness and the porosity of the biofilm, which may reduce the global efficiency of the treatment. The rotational velocity of the disk determines the quantity of oxygen available for the microorganisms in the biofilm. At low rotational velocity, the biofilm is thin and the oxygen penetrates easily through the deeper layers of the biofilm. However, as the biofilm thickens and becomes more compact, a too high rotational velocity limits the diffusion of oxygen to internal layers. This leads to anaerobic conditions in the deeper layers of the biofilm. The rotational velocity of the disk impacts the shearing forces acting on the biofilm. The shearing forces generated by the rotating disks lead to biofilm detachment, reducing the biomass and the global performance. Shearing forces are minimal at low velocity, and the biofilm is less susceptible to detaching. The rotational velocity of the disk must be carefully chosen to balance the oxygen transfer, the thickness of the biofilm and the shearing forces in order to reach an optimal performance. As the rotational velocity increases, the power required for rotating the disks increases, which leads to higher energy costs and mechanical wear. The increase in the rotational velocity of the disks beyond the recommended range also leads to an increase in noise and vibrations in the system, which has a negative impact on the operator's environment and safety. Overall, though the increase in the rotational velocity of the disk improves the treatment efficiency and the removal of organic material, there is a limit beyond which additional increases lead to lower performances, an increase in energy consumption and other negative impacts on the system. The recommended rotational velocity of the disk ranges between 2 and 5 rpm.

This parameter should meet certain operating conditions for the process to reach its highest efficiency.

The relation between the applied volume load and the rotational velocity of disks is:

$$Cv = \frac{6.37}{5.733 - r\,D}$$

where Cv is the applied volume load ($m^3.m^{-3}.h^{-1}$), r is the rotational velocity of disks ($rev.mn^{-1}$) and D is the diameter of disks (m).

Table 4.4 summarizes the values of r and D, which yield the applicable volume load.

D (m)	1	1.5	2.0	2.5	3.0
r ($rev.mn^{-1}$)	4.10	2.73	2.05	1.64	1.36

Table 4.4. *Rotational velocities and disk diameters*

The peripheral rotational velocity is given by the relation: vp = 3.14 r D.e.

The mass of a drum of 200 disks is about 10 tons in the presence of cultures, requiring a robust material and significant energy (for the rotation).

4.4.5. *Influence of temperature*

Temperature has a significant effect on the performances of disk bioreactors. In general, when temperature increases, the biological activity and the growth of microorganisms on the biofilm also increase, which is reflected by higher treatment efficiency. However, the optimal range of temperatures for the performances of RBCs depends on the specific microorganisms of the biofilm and on wastewater characteristics. At low temperature, the biological activity of microorganisms of the biofilm is reduced, which leads to a lower treatment efficiency. A too high temperature (> 35°C) is detrimental to the performances, as it destroys the structure of the biofilm and removes microorganisms from disks. For most RBC systems, the optimal temperature ranges between 20 and 30°C. Similar to all biological phenomena, the effect of temperature is an important parameter for the RBC-based treatment. It can be admitted that the operating temperature is equal to that of water, as disks pass alternatively from the air phase to the water phase.

There is an evolution of the rate of growth (μ) of bacteria with temperature according to this relation:

$$\mu_T = \mu_{10}\, \alpha^{(T-10)}$$

4.4.6. *Influence of pH*

A pH ranging between 6 and 8 corresponds to better treatment conditions. However, similar to biofilms of biofilters and hybrid systems, biomass is not sensitive to pH variations in this range of values.

4.4.7. *Influence of fats (oil and grease)*

The oxygen diffusion process is limited, even zero, when on the turf surface there is a film of fat organic substances, facilitating the formation of anaerobic zones. An oil removal tank at the head of the system is therefore essential for obtaining proper treatment conditions and maintaining homogeneity in the microbial flora.

4.4.8. *Land use*

A global surface of 1 to 5 m^2/E.H. should be considered. This system is adapted for small and average collectivities with pollution loads ranging between 300 and 2,000 I.E. (inhabitant equivalent).

4.5. Advantages and drawbacks of the system

4.5.1. *Advantages*

– Low power consumption (1 kWh.kg^{-1} removed BOD_5).

– Simple use.

– Well-thickened sludge in the settling-digesting tank.

– Good resistance to temporary organic and hydraulic overloads.

– Sludge management facilitated through combination with reed beds.

4.5.2. *Drawbacks*

– Need for staff skilled in electro-mechanics (but clearly increasing reliability).

– Limited reduction of nitrogen in classical dimensioning.

– Sensitivity to prolonged power outages that lead to unequal development of the biological film between the risen and the immersed parts (desiccation of the risen part during the stop).

In time, solid materials and the biofilm accumulate on the rotating disks, reducing the surface available for microorganisms and impacting the treatment efficiency. This leads to clogging, fouling and higher maintenance needs. Regular cleaning and maintenance of the bioreactor are required to avoid the accumulation of solids. When cleaning is insufficient despite all implemented solutions, the replacement of rotary disks allows maintaining the surface available for microorganisms and guarantees water qualities.

Compared to other biological treatment systems, bioreactors may be less efficient in removing certain pollutants. The combination of bioreactors with other treatment technologies, such as activated sludge or membrane filtration, may help improve the removal of specific pollutants.

New materials, such as graphene and carbon nanotubes, are being studied for their potential use in RBCs. This technology can be associated with tertiary treatments of removal of micro-pollutants (ozone, active carbon), membrane filtration and disinfection (chlorine, UV, ozone) in order to reuse wastewaters.

4.6. RBC processes

4.6.1. *Various suppliers*

There are many suppliers of this technology on the market. Table 4.5 presents some of them.

Company	Trademark
Veolia Environnement	Ecodisk®
A.E.I.C with SAUR, EAU PURE, E.R.E, Lyonnaise des Eaux, VALBIO	S&P RBCs ® (Germany)
Degrémont, R&O	MITA Biorulli RBCs ® (Italy)
Lyonnaise des Eaux	KEE RBCs ® (Great Britain)
NEVE Environnement	Immersed contactors Shannon® (Great Britain)
Degrémont	Klargester RBCs ® (Ireland)
ODIPURE	MITA RBCs ® (Italy)
EXELIO	PAMCO RBCs ® (Belgium)
AQUACORP	S&P RBCs ® (Germany)
PRODALL	Klargester RBCs ® (Ireland)

Table 4.5. *RBC suppliers*

4.6.2. *Focus on the Ecodisk® process (Veolia)*

This treatment system includes a compact pretreatment system allowing the removal of coarse waste, sands and greases before the main treatment in the RBCs.

At the outlet of the RBCs, the effluent can be directed towards a clarifier, reed beds or filtering screens (Hydrotech™ type). The sludge recovered in the system can feed reed beds, drying beds or any other means of mechanical dehydration. Bed supply is automatically controlled (automated valves) in order to alternate treatment periods and sludge drying periods.

In order to comply with increasingly stricter standards in terms of nitrogen effluent, the Ecodisk® process integrates nitrification and denitrification stages that maintain a TN level below 20 $mg.L^{-1}$.

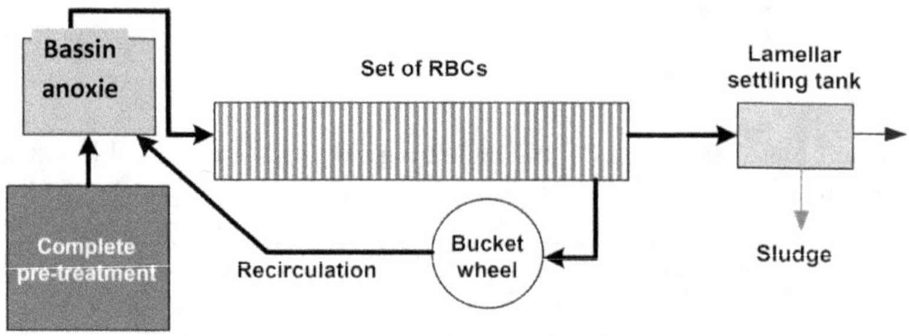

Figure 4.8. *Ecodisk® system with denitrification stage*

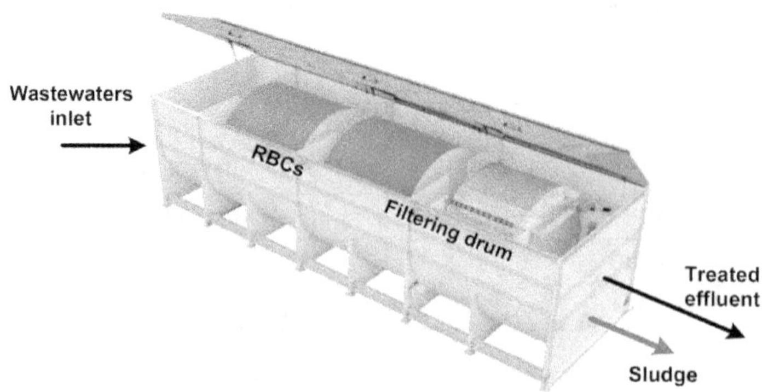

Figure 4.9. *Ecodisk® Filter process*

The combination of the Ecodisk® process and drum filtration is implemented in Ecodisk® Filter. This solution brings together in the same tank the RBCs and the drum filter, mounted on the same shaft. The filter contains thin mesh panels fixed on a rotating drum. The effluent flows under gravity through the filter, from inside to outside the drum. Suspended solids contained in the effluent are retained by the panels inside the drum.

Filter washing is regularly conducted (backwashing with water sent from outside to inside the drum) and is ensured by a system of nozzles fed by treated water (with a pump under 7 bars pressure).

The separation of sludge from the effluent treated downstream of the RBCs can be done by lamellar clarification or by drum filtration. Lamellar clarification is less sophisticated and has the advantage of simple use but has to be regular.

Figure 4.10. *Ecodisk® process facilities*

4.7. References

Alexandre, O., Boutin, C., Duchène, P., Lagrange, C., Lakel, A., Liénard, A., Orditz, D. (1998). Filières d'épuration adaptées aux petites collectivités. Technical Document, FNDAE Cemagref Editions, Montpellier.

Baillard, G. (1983). Le lagunage activé par biodisques. *L'Eau, L'Industrie, Les Nuisances*, 78, 94–102.

Ball, R.O. (1982). The effects of hydraulic variation on fixed film reactor performance. In *Proceedings of the 1st International Conference on Fixed-Film Biological Processes*, Kings Island.

Bergs, M.A. (1982). The operator's viewpoint of wastewater treatment using rotating biological contactors. In *Proceedings of the 1st International Conference on Fixed-Film Biological Processes,* Kings Island.

Boiler, M., Gujer, W., Nyhuis, G. (1990). Tertiary rotating biological contactors for nitrification. *Water Sci. Technol.*, 22(1/2), 89–100.

Borghi, M.D., Palazzi, E., Parisi, F., Ferraiolo, G. (1985). Influence of process variables on the modelling and design of a rotating biological surface. *Water Res.*, 19(5), 573–580.

Boutin, P. (1976). Les disques biologiques dans systèmes de traitement biologiques par bactéries fixées. Report, École Nationale de la Santé Publique, Rennes.

Boutin, C., Caquel, O., Dimastromatteo, N., Durot, M.-A., Fernandes, G., Parotin, S., Prost Boucle, S., Rodier, A., Tscherter, C. (2015). Ouvrage de traitement par disques biologiques. Technical Document, ONEMA and EPNAC.

Bovendeur, J., Spui, J.F., Hermans, P.G.B., Klapwi, J.K.A. (1990). Mechanisms and kinetics of COD removal and simultaneous oxygen consumption in a rotating biological contactor bio-film. *Water Sci. Technol.*, 22(1/2), 75–87.

Brenner, R.C., Heidman, J.A., Opatken, E.J., Petrasek Jr., A.C. (1984). Design information of rotatory biological contactors. Technical Document, Environmental Protection Agency, Washington, D.C.

Cortez, S., Teixeira, P., Oliveira, R., Mota, M. (2008). Rotating biological contactors: A review on main factors affecting performance. *Rev. Environ. Sci. Biotechnol.*, 7, 155–172.

Del Borghi, M., Palazzi, E., Parisi, F. (1985). Influence of process variables on the modelling and design of a rotating biological surface. *Water Res.*, 19(5), 573–580.

Di Palma, L. and Verdone, N. (2009). The effect of disk rotational speed on oxygen transfer in rotating biological contactors. *Bioresour. Technol.*, 100(3), 1467–1470.

Egli, K., Fanger, U., Alvarez, P.J., Siegrist, H., van der Meer, J.R., Zehnder, A.J. (2001). Enrichment and characterization of an anammox bacterium from a rotating biological contactor treating ammonium-rich leachate. *Arch. Microbiol.*, 175, 198–207.

Friedman, A.A., Robbins, L.E., Woods, R.C. (1979). Effect of disk rotational speed on rotating biological contactor efficiency. *J. WPCF*, 51(11), 2678–2690.

Gervasi, C. and Pascal Molle, P. (2012). Association de disques biologiques et de lits de clarification-séchage plantés de roseaux. Technical Document, Office français de la biodiversité [Online]. Available at: https://oai-gem.ofb.fr/exl-php/vue-consult/ofb_recherche_oai/DOC00083194.

Griffin, P. and Findlay, G. (2000). Process and engineering improvements to rotating biological contactor design. *Water Sci. Technol.*, 41(1),137–144.

Gujer, W. and Boiler, M. (1990). A mathematical model for rotating biological contactors. *Water Sci. Technol.*, 22(1/2), 53–73.

Kinner, N.E., Curds, C.R., Meeker, L.D. (1988). Protozoa and metazoa as indicators of effluent quality in rotating biological contactors. *Water Sci. Technol.*, 20(11/12), 199–204.

Les services de l'État dans le Gers (2021). Evaluation des Procédés Nouveaux d'assainissement des petites et moyennes collectivités (EPNAC). Technical Document, EauFrance and EPNAC.

Liénard, A. and Bardeau, L. (1993). Epuration des eaux usées par disques biologiques : étude bibliographique, tome 1. Research Report, IRSTEA, Aix-en-Provence.

Masuda, S., Watanabe, Y., Ishiguro, M. (1991). Biofilm properties and denitrification in aerobic rotating biological contactors. *Water Sci. Technol.*, 23(7/9), 1355–1363.

Mizyed, A. (2021). Review on application of rotating biological contactor in removal of various pollutants from effluent. *Tech. Biochem. Med.*, 2(1), 41–61.

Nyhuis, G. (1990). Enlargement of sewage treatment plants for nitrification by use of subsequent rotating biological contactors. *Water Sci. Technol.*, 22(3/4), 161–168.

Patwardhan, A. (2003). Rotating biological contactors: A review. *Ind. Eng. Chem. Res.*, 42, 2035–2051.

Reh, C.W., Wilson, T.E., Srinivasaraghavan, R. (1977). An approach to design of RBCs for treatment of municipal wastewater. In *ASCE National Environmental Engineering Conference*.

Rusten, B. and Odegaard, H. (1986). Kinetics of denitrification in rotating biological contactors using municipal sewage as carbon source. *Norwegian Water Technology Centre AIS*, 42(3), 194–200.

Tanaka, K., Oshima, A., Rittman, B.E. (1986). Performance evaluation of rotating biological contactor process. *Water Sci. Technol.*, 19(3/4), 483–494.

Vasiliadou, I., Molina, R., Martínez, F., Melero, J. (2014). Experimental and modeling study on removal of pharmaceutically active compounds in rotating biological contactors. *J. Hazard. Mater.*, 274, 473–482.

Wang, K., Wang, M.-H., Poon, C.P.C. (1984). Rotating biological contactors. *Effluent Water Treatment J.*, 24(3), 93–97.

Waqas, S. and Bilad, M.R. (2019). A review on rotating biological contactors. *Indonesian J. Sci. Technol.*, 4(2), 241–256.

Waqas, S., Harun, N.Y., Sambudi, N.S., Bilad, M.R., Abioye, K.J., Ali, A., Abdulrahman, A. (2023). Review of rotatory biological disks for wastewater treatment. *Water*, 15(10), 1913–1925.

Wilson, F. (1993). Kinetics and reaction order in rotating biological contactors using TOC. *Water Res.*, 27(5), 1423–1429.

Wilson, R.W., Murphy, K.L, Stephenson, J.P. (1980). Scale-up in rotating biological contactor design. *Water Pollution Control Federation*, 52(3), 610–621.

Ziembinska-Buczynska, A., Ciesielski, S., Zabczynski, S., Cema, G. (2019). Bacterial community structure in rotating biological contactor treating coke wastewater in relation to medium com-position. *Environ. Sci. Pollut. Res.*, 26(19), 19171–19179.

Index

L, M, O

P, R

S, T, U

Other titles from

ISTE

in

Ecological Science

2026

GAID Kader
Wastewater Engineering and Treatment 1: Composition, Pollutant Loads, Primary Settling, Activated Sludge

2025

COURANT Dimitri, REBER Bernard
Deliberative Democracy and Ecological Transition: The French Citizens' Convention for Climate
(Planetary Changes, Decision, Governance Set – Volume 1)

2024

AMIARD Jean-Claude, ZERBIB Jean-Claude
Radioactive Risk for Humans
(Radioactive Risk Set – Volume 7)

AKBAR Saima, KUMAR SINGH Suraj, KANGA Shruti
Disaster Management: Preparedness and Mitigation

FLEURENCE Joël
Aquatic Biotechnologies: From Genetic Engineering to Enzymatic or Fermentation Engineering

GAID Kader
Wastewater Reuse 1: Characteristics, Uses, Applications, Filtration and Disinfection of Water

GAID Kader
Wastewater Reuse 2: Micropollutants, Membranes and Treatment Procedures

MATAGNE Patrick
Birth of Scientific Ecology: Eugenius Warming (1841–1924)

2023

GAID Kader
Drinking Water Treatment 1: Water Quality and Clarification
Drinking Water Treatment 2: Chemical and Physical Elimination of Organic Substances and Particles
Drinking Water Treatment 3: Organic and Mineral Micropolluants
Drinking Water Treatment 4: Membranes Applied to Drinking Water and Desalination
Drinking Water Treatment 5: Calco-carbonic Equilibrium and Disinfection

2022

AMIARD Jean-Claude
Marine Radioecology
(Radioactive Risk Set – Volume 6)

GODET Laurent, DUFOUR Simon, ROLLET Anne-Julia
The Baseline Concept in Biodiversity Conservation: Being Nostalgic or Not in the Anthropocene Era

GRISON Claude, CASES Lucie, LE MOIGNE Mailys, HOSSAERT-MCKEY Martine
Photovoltaism, Agriculture and Ecology: From Agrivoltaism to Ecovoltaism

LÉVÊQUE Christian
Biodiversity Erosion: Issues and Questions

2021

AMIARD Jean-Claude
Disarmament and Decommissioning in the Nuclear Domain (Radioactive Risk Set – Volume 4)
Management of Radioactive Waste (Radioactive Risk Set – Volume 5)

FLEURENCE Joël
Microalgae: From Future Food to Cellular Factory

JOLY Fernand, BOURRIÉ Guilhem
Mankind and Deserts 1: Deserts, Aridity, Exploration and Conquests
Mankind and Deserts 2: Water and Salts
Mankind and Deserts 3: Wind in Deserts and Civilizations

PIOCH Sylvain, SOUCHE Jean-Claude
Eco-design of Marine Infrastructures: Towards Ecologically-informed Coastal and Ocean Development

2020

BRUSLÉ Jacques, QUIGNARD Jean-Pierre
Fish Behavior 1: Eco-ethology
Fish Behavior 2: Ethophysiology

ROSSIGNOL Jean-Yves
Climatic Impact of Activities: Methodological Guide for Analysis and Action

2019

AMIARD Jean-Claude
Industrial and Medical Nuclear Accidents: Environmental, Ecological, Health and Socio-economic Consequences
(Radioactive Risk Set – Volume 2)
Nuclear Accidents: Prevention and Management of an Accidental Crisis
(Radioactive Risk Set – Volume 3)

BOULEAU Gabrielle
Politicization of Ecological Issues: From Environmental Forms to Environmental Motives

GIRAULT Yves
UNESCO Global Geoparks: Tension Between Territorial Development and Heritage Enhancement

KARA Mohamed Hichem, QUIGNARD Jean-Pierre
Fishes in Lagoons and Estuaries in the Mediterranean 2: Sedentary Fish
Fishes in Lagoons and Estuaries in the Mediterranean 3A: Migratory Fish
Fishes in Lagoons and Estuaries in the Mediterranean 3B: Migratory Fish

OUVRARD Benjamin, STENGER Anne
Incentives and Environmental Policies: From Theory to Empirical Novelties

2018

AMIARD Jean-Claude
Military Nuclear Accidents: Environmental, Ecological, Health and Socio-economic Consequences
(Radioactive Risk Set – Volume 1)

FLIPO Fabrice
The Coming Authoritarian Ecology

GUILLOUX Bleuenn
Marine Genetic Resources, R&D and the Law 1: Complex Objects of Use

KARA Mohamed Hichem, QUIGNARD Jean-Pierre
Fishes in Lagoons and Estuaries in the Mediterranean 1: Diversity, Bioecology and Exploitation

2016

BAGNÈRES Anne-Geneviève, HOSSAERT-MCKEY Martine
Chemical Ecology

2014

DE LARMINAT Philippe
Climate Change: Identification and Projections

www.ingramcontent.com/pod-product-compliance
Lightning Source LLC
LaVergne TN
LVHW020458140526
838656LV00001B/30
* 9 7 8 1 8 3 6 6 9 0 5 6 6 *